Dalit Ecologies

Dalit Ecologies explores the ecological experiences, histories, and perspectives integrated within Dalit writing, art, and culture. Aligning with theories of environment justice and ecological struggles experienced by black populations, the book delves into six major themes: caste, earth, and earthly environment; labour and mobility; casteization of technology and industry; climate justice; Dalitbahujan Anthropocene; and eco-literary tradition. The essays in the book include the relationship between caste and environment, Dalit autobiographies, folktales and novels, city, waste and discard, caste-based industry and occupation, technological injustice, weather, caste and climate change, and black–Dalit ecologies. Expanding the boundaries of environmental studies, the book brings attention to individuals such as Adwaita Mallabarman, Bama, Nek Chand, and Deena-Bhadri on the one hand and specific places and arenas like the rock garden, tannery, brick kiln, steel industry, and sanitation on the other, which are often overlooked in mainstream environmental writings.

Mukul Sharma is a professor of environmental studies at Ashoka University, India. He has published *Caste and Nature: Dalits and Indian Environmental Politics* (2017), *Defining Dignity: An Anthology of Dreams, Hopes and Struggles* (ed., 2005), and *Unquiet Worlds: Dalit Voices and Visions* (ed., 2004). He has worked in media and civil society organizations and was associated with World Social Forum, World Dignity Forum, and National Confederation of Dalit Organizations.

Dalit Ecologies

Caste and Environment Justice

Mukul Sharma

CAMBRIDGE
UNIVERSITY PRESS

Shaftesbury Road, Cambridge CB2 8EA, United Kingdom

One Liberty Plaza, 20th Floor, New York, NY 10006, USA

477 Williamstown Road, Port Melbourne, VIC 3207, Australia

314–321, 3rd Floor, Plot No. 3, Splendor Forum, Jasola District Centre, New Delhi – 110025, India

103 Penang Road, #05–06/07, Visioncrest Commercial, Singapore 238467

Cambridge University Press is part of Cambridge University Press & Assessment, a department of the University of Cambridge.

We share the University's mission to contribute to society through the pursuit of education, learning and research at the highest international levels of excellence.

www.cambridge.org
Information on this title: www.cambridge.org/9781009453455

© Mukul Sharma 2024

First published 2024

Printed in India by Avantika Printers Pvt. Ltd.

A catalogue record for this publication is available from the British Library

ISBN 978-1-009-45345-5 Hardback

For Ravi Agrawal, a distinguished environmental author, artist, photographer, and curator, whose unwavering passion, encouragement and belief in my work has infused me with energy and confidence needed to persevere in my research and writing

and

For the esteemed Hindi writer Ramesh Upadhyay and his literary endeavour Kathan, *whose nurturing support played a pivotal role in shaping the foundational aspects of my thoughts*

Contents

Foreword

I remain hopeful as scholars like Mukul Sharma add a critical perspective to environmental scholarship, specifically race, ethnicity, and class. His analysis of Dalits, untouchables, the lowest caste in India is traced throughout *Dalit Ecologies: Caste and Environmental Justice*. Sharma is part of growing and critical international scholarship and literature that focuses on marginalized persons of colour and the environment. The value of Sharma's work rests on secondary sources and primary sources with the primacy of voices of people who are both depressed and creative. Dalits tell their own stories and create ecologies. The oppressors are just that – oppressors – and not central to the narrative.

Sharma's intriguing book takes me back to my days as a PhD student in history in the 1990s. During that time, I was excited and afraid, quickly learning there was very little diversity in environmental history. With my dissertation proposal, I did the opposite of what dissertation advisors advise: write your dissertation based on an established area and just get it done. Willfully, I ignored the advice and plunged into the unknown of nascent African American environmental history. I did not know if archival sources, critical to writing such a history, would be fruitful. Little or racist mention of African Americans and the environment stood between my research and completion of my dissertation.

In the 1980s and 1990s, many white scholars marginalized and objectified Native Americans and African Americans. Some environmental historians focused on the tensions and comparisons between Native Americans and white voices. Still the narratives of whites and their voices remained at the fore, essentially ecological imperialism, what was done by whites to Native Americans, including spreading disease from Europe to the Americas and Caribbean. Agricultural history seemed a logical path, particularly the scholarship of the American South. Yet African Americans were objects and not the voices from primary sources. Theirs were supporting and cameo roles in relationship to whites and agricultural equipment.

Early in my academic career, I often found myself alone and in environmental history circles the subject matter of African American environmental history was objectified, rejected, or sidelined. I was often asked at conferences, 'Where were the white people?' Looking back, there was sarcasm in my tone when I said, 'In your books,' when speaking to a mostly male and white audience.

My mentor Mart V. Stewart pushed against this dominant white narrative in his book, *What Nature Suffers to Groe: Life, Labor, and Landscape on the Georgia Coast, 1680–1920* (1996). He was my first conversation partner in African American environmental history and the beginnings of a broader environmental scholarly circle for me. From there, I forged a cohort of white environmental scholars that resulted in my edited volume *'To Love the Wind and the Rain': African Americans and Environmental History* (2006). African American authors followed with books focusing on African Americans and the environment, including Carolyn Finney's groundbreaking *Black Faces, White Spaces: Reimagining the Relationship of African Americans to the Great Outdoors* (2014).

Equally as important as this scholarship is remembering the contributions of black historical figures, the ancestors to black environmentalism, including slave insurrectionist Nat Turner, scientist George Washington Carver, underground railroad conductor Harriet Tubman, and environmentalist Wangari Maathai. Environmental histories can be based on first-hand accounts, including Drew Lanham's celebrated environmental autobiography *The Home Place: Memoirs of a Colored Man's Love Affair with Nature* (2016). The same is true for *Dalit Ecologies*.

One bridge between this literature of people of African descent and the environment and Dr Mukul Sharma's scholarship on Dalits and environmental justice is Isabel Wilkerson's *Caste: The Origins of Our Discontent* (2020). Wilkerson's non-fiction compares racism in the United States to caste systems in Nazi Germany and India.

For readers in the United States, Wilkerson's work serves as an on-ramp to Sharma's groundbreaking, important, and critical analysis of Dalit ecologies. Sharma highlights the experiences of being on the lowest rung of the crushing caste system, and its environmental implications. As was true of my experience as a PhD student, Sharma is one of the first to focus on Dalits on the fringes, those who are outcastes, through the lens of environmentalism. If not for Sharma, the environmental voices of Dalits would otherwise be muted and invisible. I feel a kindred spirit with Sharma as we share new beginnings for African Americans and Dalits, both subjugated yet hope-filled people. He does so by grounding his work in 'fieldwork and visits, interviews and conversations with Dalits'. He gives a voice and space to people in the lower social-economic strata with stories of Mali

Nek Chand who built a rock garden from waste and also a storyteller on behalf of Dalits. Both Sharma and Chand excavate expressive and deeply meaningful environmental justice stories, including that which puts a mirror up to simple to grand human experiences.

The stories of people who are marginalized because of ethnicity, society, and class continues.

Dianne D. Glave
Author of *Rooted in the Earth: Reclaiming*
the African American Environmental Heritage (2010)

Acknowledgements

This book emerged as a response to an array of insightful comments, thought-provoking questions, and constructive reviews that were shared during various individual seminars, talks, conferences, and forums, as well as in classrooms, subsequent to the publication of my previous book *Caste and Nature: Dalits and India Environmental Politics* in 2017 and its Hindi edition in 2020. Throughout these engagements, a multitude of topics were explored, generating curiosity and stimulating discussions. These encompassed subjects such as contemporary caste dynamics, the experiences of Dalits, the environment, technology, industrialization, urban environments, the Anthropocene epoch, the concept of commons, the interplay between religion and environmentalism, and the intersection of literature and culture with these issues. Of particular interest to the participants was the envisioning of the future trajectory of Dalit environmentalism. This book delves deeper into these areas of inquiry, expanding upon the themes and insights initiated by my earlier work.

I would like to express my sincere gratitude to all those who have played a pivotal role in bringing this present work to fruition. I am indebted, first and foremost, to the Ambedkar Study Centre at Seshadripuram Evening Degree College in Bengaluru. I extend my great appreciation to N. S. Satish, the principal of the college, and Dhareppa Konnur, the programme coordinator. Their encouragement and support prompted me to initiate an international webinar series in 2021 on 'Anti-Caste Politics and Environmental Justice', which proved to be a catalyst for several ideas that shaped this book. I would also like to extend my thanks to the Antiracist Research and Policy Centre (ARPC) at the American University in Washington, DC, USA. Dr Malini Ranganathan, in particular, played an instrumental role in collaborating with the webinar series, and her valuable contributions have greatly enriched this work. I am thankful to Ravi Agarwal and his initiatives through the Goethe Institut and Max Muller Bhavan. It was through these platforms that I had the opportunity to present and engage in discussions on various topics addressed in this book. Ravi and I maintained a constant exchange of ideas and explored avenues for collaboration on this subject, which proved to be immensely stimulating.

I would like to express my utmost gratitude to my colleagues and friends who have been a part of various conferences and seminars. Our interactions during these events have been invaluable as we shared thoughts, exchanged materials, and engaged in fruitful discussions. Christopher Coggins, Ambika Aiyadurai, Eliza Kent, Ujjwal Singh, Awadhendra Sharan, Hilal Ahmed, Abhay Kumar Dubey, Aditya Nigam, Suraj Yengde, Joya John, Ranjit Hoskote, Amruta Nemivant, Vidya Shivadas, Vandana Singh, Sahil Nijhawan, Vidyadhar Atkore, Suhas Bhasme, Mousumi Ghosh, Samira Agnihotri, Ravivarma Kumar, Arul Mozhi, J. Srinivasan, Kenneth Bo Nielson, Thomas Crowley, Goldy M. George, Assa Doron, Srilata Sircar, Sidharth Ravi, and Prakash Kashwan.

I extend my sincere thanks to numerous Dalit activists, writers, and friends who have graciously shared their invaluable insights with me. Among them, I express my special appreciation to Ashok Bharti from the National Confederation of Dalit Organizations (NACDOR), Paul Divakar from the National Campaign on Dalit Human Rights (NCDHR), Basudev Sunani, Savi Sawarkar, and Prashant Tambe. Their perspectives and knowledge have played a crucial role in shaping this work, and I am deeply indebted to them for their contributions.

Additionally, I would like to acknowledge the profound influence of several African American and black academicians, who have enriched my understanding of related subjects. Their contributions have greatly enhanced the depth and quality of this work. In particular, I am extremely grateful to Dianne D. Glave and James Roane for their significant contributions to my research. I have had the privilege of occasionally sharing my work with James Roane, and I have found his writings to be highly relevant and thought provoking.

I would like to express my appreciation to my colleagues at Ashoka University, who have consistently provided support for my research. Thanks to Upinder Singh, Mahesh Rangarajan, Aniket Aga, and Divya Karnad for their invaluable assistance and encouragement. The students at Ashoka University, who have enrolled in my course, have played a significant role in shaping this book. Although it is challenging to name them individually due to their large numbers, I am immensely grateful for their contributions to the development of this work.

I also extend my special thanks to Prasenjit Duara, Amitav Ghosh, and K. Sivaramakrishnan. Though our interactions may have been intermittent, their responses have been instrumental in refining this book.

There are no words sufficient to express my gratitude towards the friends who have been steadfast in their support throughout the completion of this book. I convey my warm thanks to Anand Swamy, Sumedha Basu, Anand Pradhan,

Aseem Shrivastava, Jeremy Seabrook, Maninder Thakur, Ravikant, and Praveen Jha for their unwavering presence and contributions. Anwesha Rana from Cambridge University Press has supported the idea of this book since its inception. Anonymous readers appointed by the Press offered some astute comments that provoked good revisions in the manuscript.

Some essays of this book have been published in an earlier form in the *South Asia*, *Social Change*, *South Asian History and Culture*, *Environment and Society* and *Man in India*.

I am deeply grateful to my family members, Charu, Ishaan, my mother Shashikala Sharma, niece Appi, and B. P. Singh, as well as numerous others, for their unwavering support and warmth throughout the process of completing this book. During the challenging period of enduring long COVID times, their tremendous support played an instrumental role in helping me navigate through the difficulties. Without their presence and encouragement, I would not have been able to overcome the obstacles and bring this book to completion.

1

Introduction

This book comes out of my continuing research on the interrelationship between caste, Dalits, and the environment.[1] The terminology of *Dalit Ecologies*[2] is an open, collaborative space to collate insights from Dalit, backward, and anti-caste histories, socio-political thought, and movements and organizations, and to bring to light some aspects of contemporary Dalit environmental narratives. My deep interest in the subject inspired me to introduce an elective course on 'Environment and Social Exclusion' at Ashoka University in 2020, which proved to be very popular, and initiate an international webinar series on 'Anti-Caste Politics and Environmental Justice' and 'Dalit Ecologies' in 2020–21, organized by the Ambedkar Study Centre, Seshadripuram Evening Degree College, Bengaluru, in association with the Antiracist Research and Policy Centre, American University, Washington, DC.[3]

My conception of *Dalit Ecologies* is also influenced by, and derives from, rich academic and political work in other parts of the world around 'Black ecologies', 'Racial ecologies', and 'Feminist ecologies', which analyse the interrelation between race, gender, and environmentalism through an interdisciplinary lens of race and gender studies, and socio-political ecology.[4] Simultaneously, they sculpt distinct histories, stories, and approaches of racialized communities in expressing ecological relations. I draw insights from collective thinking and initiatives on 'Global Black Ecologies',[5] which focus on 'the critical insights, world-making, and world-sustaining practices of global Black communities'. While primarily Afro-diasporic in nature, they also express 'connections to communities without recent African ancestry who have articulated political and social connections to Blackness with respect to their position within racialized national and global hierarchies, such as caste, exemplifying Achille Mbembe's conceptualization of *"Becoming Black of the World"*.[6]

The chapters are grounded in my fieldwork and visits, interviews, and conversations with Dalit, backward caste, and anti-caste people and activists, grassroots organizations and NGOs involved in Dalit issues, leaders of political parties, industrialists, and government officials – in the states of Bihar, Uttar Pradesh, Delhi, Haryana, and Punjab.[7] Dalit autobiographies, coming out from the states of Tamil Nadu, Karnataka, Maharashtra, and West Bengal, have also been insinuated to comprehend the wide canvass of the subject. The chapters here include rural and urban environments; industrial, technological, and agricultural systems; developmental and anthropogenic activities; economic, political, and cultural spheres; and Dalit social and political organizations. They offer a range of perspectives on how, and in what ways, Dalit ecologies can be constituted, which also signify particular identities and ecological places and spaces. For example, Musahars, an 'untouchable' caste in the Indo-Gangetic Plain of India and Nepal, create their ecological world through folktales, stories, and songs, woven around the two warrior brothers Dina and Bhadri. Musahars situate them as their ecological ancestors in contemporary struggles over lands, ponds, and rivers, particularly during the agricultural seasons in the states of Bihar and Uttar Pradesh. Nek Chand, a low-caste Mali (a caste that has been described as embodying nurturers and gardeners of the earth), builds a Rock Garden to celebrate peoples' lives and document his story of Anthropocene through wasted, discarded, and broken materials, focusing on a history of violence in the making of the modern Chandigarh city in the state of Punjab. Paul Divakar, a key organizer of National Campaign on Dalit Human Rights, says that the daily existence of Dalits – their labour, place, occupation, discrimination, violence, and the intensity of living in hard and hostile situations – are the real issues of Dalit climate environmentalism. Each of the subjugated castes – the lower, the lowest, the untouchables (or ex-untouchables) – can have their own vantage points to underline the ecological consequences of the caste system, and its enduring old and new lives in Indian society. There are multiple Dalit ecologies, which result in a collage, creating a new whole. Diverse perspectives cross through a common juncture, that is, domination of Hindu Brahmanical attitudes and practices, and their control and influence over the natural–physical world. The Brahmanical world embodies binaries of Dalits and others, untouchables and *savarnas*. The marginalization and stigmatization of Dalits based on caste–property–production relations are highly evident. Equally importantly, the dominance of castes in control of natural and human resources, as well as industrial and cultural power, plays significant roles in shaping environmental politics. However, Dalit ecologies commonly epitomize a recurrent migration of communities between two different worlds – one of everyday ecological burdens in a marked hierarchal order and another of their indigenous imaginations, particularly in relation to earth, life, landscape, place, animal, waste, and work.

Dalit Ecologies appears at the intersection of three critical distinctions of the twenty-first century: One is an increased academic and political interrogation of manifold dimensions of caste power, by which caste appropriation, subjugation, and domination is produced and developed in every sphere of society. Second is a globalization of theory and practice of the environmental justice movement, which goes beyond a focus on differential toxic exposure of African Americans, and makes it politically potent internationally, impacting the mainstream environmental agenda. And third is a phenomenal rise of Dalit and backward caste political and cultural power, which is simultaneously marked by differences and complexities of regions, local histories, ideologies, interests, and classes. However, their movements are as much for articulating their position on space, environment, natural, and physical resources as they are for achieving political representation, identity, and development. This collection captures some key actions, writings, and movements of Dalit individuals and organizations who grapple with issues of Anthropocene, climate change, waste, urbanization, industry, technology, and science, demonstrating a range of issues appearing under the broad scope of *Dalit Ecologies*. They show the long distance travelled by Dalits in thinking and activism as they increasingly address environmental and social injustices, and contemporary environmental challenges.

Dalit Ecologies is also located at the cusp of deep tensions built around the interrelationship between anti-caste movements and environmental movements, Dalitism and environmentalism.[8] There are several challenges in dealing with these two streams together under *Dalit Ecologies*. The main aim of ecology is to understand the environment of humans and non-humans, living and non-living, biotic and abiotic, and their interaction with each other. Yet Dalit and anti-caste movements, in diverse ecologies, have been historically confronted with social hierarchy and authority of power, and have fought for dignity, justice, and equality. Centuries of caste- and birth-based subjugation, denial of human existence, and a search for their place on earth have made Dalit visions anthropocentric. Concepts of biosphere and biodiversity of organisms, and one or common future, have been questioned by Dalits, pushing them to address caste social injustices 'here' and 'now'. It is argued that in the name of greater good, Dalit present and future should not be asked to bear the burden of caste, and capitalist exploitation of the environment in the past.[9] I take such tension points in Dalit political articulation as a critical source to reflect on their different attitudes to, and practices of, imagining environmental relations amidst exploitative social and economic structures. In my perspective, a low-caste Mali Nek Chand, who builds a Rock Garden from waste in a modern Indian city, is seen as a collector and chronicler of planet and people, as he establishes distinct and different everyday relations with urban waste and natural rock, in times of 'great acceleration' in the country.

Dalit Ecologies is 'hybrid'[10] in nature, where different regimes of production of nature exist simultaneously. The book thus talks about development and its close companions – industry, technology, science, and modernity. The lives of Dalits and low-caste people have been changing under industrial, technological, and urban development, which comes with increased politicization, mobilization, migration, and communication. However, these have not dissolved unequal caste relations in work, space, and the environment. Leather tanneries in Kanpur, Uttar Pradesh, from the eighteenth to twentieth centuries, have been progressing mainly through particular manual applications of Dalits, harnessing their physical labour within the factory, facilitated through caste economy. Technology and caste function together – occupational and otherwise – where a Dalit has to be a 'Dalit'. While Dalits increasingly migrate to the city and work in modern 'polluted' industrial set-ups, there are glaring gaps between industrial/urban development and caste/Dalit occupation. *Dalit Ecologies* demonstrates the divide between the universal promises of development and its ground realities in particular cultural and caste locations. However, there are many examples of how new places, industries, technologies, and organizations can be used to further Dalit labour and environmental agendas. Despite various complexities and ambiguities, where environment cannot be separated from the project of caste and capitalist hegemony, Dalits show the paths of building a Dalit environmentalism and visualizing a Dalit–low-caste environment justice, based on the rights of Dalits that affirms their material and cultural claims in nature.

In *Dalit Ecologies*, political economy and power appears prominently to bear on environmental inequalities and struggles against them. The perspectives of Dalit individuals, politics, ideologies, and organizations on environmental change and stories around ecology are deeply entangled with differential histories and presents of access to natural resources. In the river Titash,

Dalit Malo fisherfolk give moving accounts of histories of nature, in conjunction with culture and the human and non-human world, where modalities of power, reflected in caste, class, and gender, determine the course of their life and labour. In evoking Dina-Bhadri, and fighting over access to land and waterbodies, Dalit Musahars ask what an environmental struggle is. Is it claims to ownership of natural resources over another as an assertion of their power? Does it also imply baseline conservation of resources, along with security of their livelihood? Is it a process of culturalization and conscientization of their identity, history, and culture? Does it signify gaining power by the oppressed? Does it hold some distinct connotation for nature and Dalits?

Dalit Ecologies attempts to take the understanding of the politics of caste, nature, and Dalits two steps forward by first locating environmental politics in Dalit literature, culture, industry, technology, Anthropocene, and justice; and second, illuminating Dalit environmental visions and voices in familiar and unfamiliar terrains. My earlier work focused more on a critique of 'eco-casteism'. Taking it ahead, this work highlights strands of caste and Dalit environmentalism, also taking comparative notes on race and the emergence of African American and black environmentalism in other parts of the world.[11] In the following sections, I continue my discussion of Dalit ecologies through six conceptual terrains: environment justice, earth and earthly environment, labour and mobility, casteization of technology and industry, and climate justice and Dalitbahujan Anthropocene.

Caste and Environment Justice

The terms 'environment racism', 'environment justice' and 'environment equality' have been articulated widely since the 1990s. In comparison, research on environment justice, caste, and Dalits has been less, and environment movements in India have not focused much on environmental injustices along caste and Dalit lines. There is ample research now to show the crucial role of caste in environment and the economy, and how Dalits are disproportionately affected in casteized nature. Locating the trajectories of theory and practice of environment justice, *Dalit Ecologies* makes a case for inclusion of caste as a critical lens to conceptualize environmental injustices.

The need to recognize plurality in defining environment justice has often been highlighted in the movement. How do environmental justice movements decide their goals and objectives? How can difference and diversity be given respect and recognition? Can the complex interplay of race, class, and gender give more meaning to the concept? It has been argued that diverse movements can have multiple meanings of justice – from distribution to individual and community recognition, to self-esteem and cultural assertion: 'Groups emphasize different notions of justice, on different issues, in various contexts; there is a flexible, heterogeneous and plural discourse of justice.'[12] There are at least four striking features in this notion of plurality: recognition and respect, importance of experience and expression, multiple spatialities of injustices, and the agency of the coloured people.

First, in the 'expanded set of justice discourses', recognition of many forms of degradation, devaluation, and humiliation at individual and social levels has been considered as a distinct element of justice, which should be integrated within the distributive framework. Recognition has structural elements – how, what, when, and where individuals and communities are recognized is considered as crucial for just distribution.[13] With recognition and respect comes an acute awareness of social contexts and norms that foster unequal distribution, as well as a positive acceptance of differences in groups, cultures, and symbols.[14]

Second, the experience becomes an important foundation for the organized politics of environmental justice. There is a recognition that there has been an overwhelming inclusion of racist environmental experiences, while those of others have been excluded, which have sharpened environmental inequalities. As has been remarked: 'Examining experience as the basis for knowledge and subjectivity allows a critique of alienating and oppressive situations, organization around those experiences or identities and an understanding of the creation of resistance to unwanted aspects of those experiences.'[15] Environmental experiences carry multiple contexts – social, cultural, historical, political, and spatial – which make them a concrete subject. Experience can be a reliable source of evidence. Thus, the environmental justice movement uses life narratives of people living in polluted areas, which are combined with hard facts and meticulous research, to challenge structures of environmental racism.

Third, spatial dimensions of discrimination and inequality have found a particular focus in environmental justice movements. Race has been located initially as a variable in the spatial layout of urban areas, including housing patterns, street and highway configurations, commercial development, and industrial facility siting.[16] The production of spaces of inequality has not been confined only to polluting industries and land use; it has had a more complex historical, cultural, and geographical dynamic. Spatial concerns have expanded into 'exploring the multiple spatialities of environmental justice' beyond distribution and proximity. Since the spatialities of environmental injustice have been diverse and expansive – from climate change and natural vulnerabilities to flooding, droughts, and hunger – new geographies, spaces, places, and scale have mattered.[17]

Fourth, coloured people have continued to hammer the key roles of their community organization, leadership, and agency in environmental justice movements. When more than a hundred community, religious, cultural, and social leaders – all people of colour – addressed a letter to ten prominent environmental organizations, they pointed out their 'racist and exclusionary practices', their lack of diversity, and failure to take up environmental justice issues. In the course of time, coloured people have asserted their right to speak for themselves. They have demanded a recognition of expertise and knowledge of their community organizations, and emphasized the priorities set by grassroots workers and activists: 'People from the outside should not come in and think that there is no leadership in the grassroots community. The people in the community should lead their own community and create legacy by teaching young people to be leaders.' Principles of democratic organizing have been further developed in arenas of inclusivity, bottom-up organizing, partnerships, accountability, fundraising, and addressing differences.[18]

While blacks have laid the foundation of environmental justice by questioning the dominant paradigm of environmentalism, the globalism of the movement has raised some pertinent questions about its conceptual trajectories. For example, in Latin America, researchers have recognized the value of environment justice as a framework of analysis and as a discourse of political action, for addressing crucial questions regarding the continent's problems of inequality and injustice. However, they have also identified certain practical and conceptual challenges for the region. From the lack of availability and quality of scientific data to limitations of funding, they have raised questions about differences in geographies, social contexts, and concepts, which have emerged from specific political–cultural–historical experiences. Land, agriculture, mining, oil, indigenous knowledge, and genetic resources have emerged as key themes.[19] Peoples' lives and livelihoods, stronger social justice and human rights components, mass mobilization, labour, gender, indigenous social and urban movements, and the church of liberation theology have been their salient features. Unlike the United States, the spatiality of environmental hazards on the basis of race and income do not typically appear in Latin American cities.[20] At the same time, race has been a critical element within the larger politics of environmental justice in Latin America, particularly in the context of the looming legacy of Afro-American slavery, segregation, and social hierarchies, which correlate to inequities in the distribution of environmental resources. The experience of race has been a significant factor in shaping access to natural resources, exposure to environmental and natural hazards, access to environmental benefits, and determinations of who counts in environmental policy formulation, implementation, and enforcement. Sundberg has empirically analyzed how systems of racial classification draw upon and come into being through environmental formations, meaning 'the historically specific articulations between environmental imaginaries, legal frameworks for allocating natural resources and patterns of environmental transformation driven by the broader political economic context'.[21] Latin American environmental justice discourse has thus generated fresh knowledge and action against race-organized and rationalized inequality.

In the African continent, environmental justice has been placed in the context of the troubled history of colonialism, where colonial control and use of natural resources were accompanied by Western white construction of environmental knowledge. Such colonial subjugation, on the one hand, portrayed Africans as incapable of environmental sensibility, who needed to be guided by educated, knowledgeable non-Africans. On the other, it created sufficient condition to support 'external intervention' and 'the interests of various actors in development', which were 'served by the perpetuation of orthodox views, particularly those regarding the destructive role of local inhabitants'.[22] The white imagination of wilderness, creation of wild enclaves, development of white tourism, opening of several destructive extractive industries, eviction and exclusion of local people from forest and natural areas, and curtailment of their human rights were the dominant themes of environmental injustices in this region. In this context, the term 'environmental justice' also implicitly revealed some potential limitations. Efforts to make environmental discourse more responsive to colonial history, and to the shaping of environmental policies by local and transnational elites in the post-colonial world, have become central to environment justice. In the course of time, environmental justice movements branched out in many directions that viewed 'environmental issues as deeply embedded in access to power and resources in society', and connected questions of environmental policy with problems of social injustice.[23] The South African Environmental Justice Networking Forum explained: 'Environmental justice is about social transformation directed towards meeting basic human needs and enhancing our quality of life – economic quality, health care, housing, human rights, environmental protection and democracy.'[24] Global discourses on environmental justice in Africa and Latin America reaffirm the basic requisite to fight against race and white domination to ensure justice. At the same time, the relationship between environmental injustices and colonialism, imperialism, neo-liberalism, and development have been added to sharpen the framework. This scholarship has also generated increasing knowledge about the dynamics of colonialism, state, market, and development, and the oppression of environmental rights of indigenous populations.[25]

In Indian and South Asian contexts, *Dalit Ecologies* has to be substantially recognized within the politics of environment justice. Such recognition will not only question eco-casteism and the domination of Hindu Brahmanical environmentalism but will also strengthen the pathways for distributional justice by ensuring equal access and rights of Dalits and those belonging to lower castes to natural resources, industry, technology, and development. However, as several chapters in this volume suggest, there are some important ecological (political, cosmological, psychic) factors that have profoundly and distinctly shaped Dalit environmental experiences. Anti-caste literature emphasizes that everyday practices, of what constitutes environmental activity and thinking, are structured by an archaeology of untouchability in body, contact, touch, smell, feel, belonging, work, and sociability. In Dalit relations to environment, they articulate certain notions of nature that go beyond physicality, possession, and distribution of resources as environmental justice discourses have done in the past. The language of experience, feeling, humiliation, and dignity, ubiquitously used in Dalit movements, also gets integrated into conceptions of environmental justice.

Earth and Earthly Environment

A man-made garden of rocks and wastes, Nek Chand's Rock Garden in Chandigarh city,[26] welcomes everybody to the 'earth story' – so one of the signboards says at its entrance gate. Low-caste Nek Chand Saini undertook the building of the garden on a small scale to narrate the story of the earth life systems. Academic works and media reports give an idea of the political–economic context in which this extraordinary endeavour began taking shape. At its core lay a relationship among four distinct aspects. First, an imagination of lives lived and developed in earth's history. Second, the presence of humans and their interrelationships with non-human worlds, leading to widespread changes in their life. Third, the earth's witness to destruction, displacement, and waste, and the creation of the earth's modern history. And fourth, along with local, folk, design, and architecture, evoking a utopia for the future, that is, representing land, water, agriculture, mountain, rock, human, animal, women, children, living space, and sacred as the earth's biosphere in a self-sustained, autonomous way.

To explain Nek Chand's Earth story, I draw inspiration from cultural historian Thomas Berry and African American environmental author Carl Anthony, who weave a new story about who human beings are, in relationship to the story of Earth. In his book *The Dream of the Earth*, Berry notes that one of the remarkable achievements of the twentieth century is its ability to tell the story of the universe from empirical observation, with amazing insights into the sequence of transformations that have brought into being Earth, the living world, and the human community together. According to him, Earth consists of a 'communion of subjects not a collection of objects'. Earth and its inherent powers bring forth a marvellous display of beauty in an unending profusion, which gets overwhelmingly transmitted to human existence and consciousness. Visualizing the future of Earth–human relations, he says that 'our own dreams of a more viable mode of being for ourselves and for the planet Earth can only be distant expressions of this primordial source of the universe itself in its fullest extent in space and in the long sequence of its transformations in time'.[27]

Reflecting on African American environmental history, and its relationship to the earth, Anthony states that the knowledge of the earth, and the place of human beings in it, gives a sense of identity and belonging to the black people. He narrates:

> The earth is the ground we walk on, the sea and air, the soil that nourishes us, the sphere of mortal life, the third planet in order from the sun, near the center of Milky Way galaxy. Everything that we do, or aim to do, is governed by our relationship with the earth – to its inspiration and resources, to our consciousness of its relationship to the cosmos, to our affinity with human and other-than-human life. Our knowledge and affinity with the earth, in all of its richness of life and diversity, stretches from the tiniest particles, waves and cells, to its plant forms and ecosystems, its rivers, mountains, and seas, to the majesty of our solar system, galaxies, and outer edges of the universe.[28]

In Dalit ecologies, we find many views of the earth – as a god and goddess, as a cultural symbol, as a guiding organism to their habitability and workability, and as a centre of their earthly environments.[29] The earth is personified as an important locus of their individuation, carving them as a free, working community, without the bearing of burdens of caste, and thus revealing irreducible instances of creation, belonging, and rights. As this book shows, in an urban landscape, Nek Chand represents Sanjhi Devi – Mother Earth – on many walls of the Rock Garden. She is surrounded by symbols of sun, moon, and vegetation. She is painted in various shapes and forms – human, animal, trees, and ideographic formless images. In a rural landscape, Dalit Musahars, residing in districts of Bihar and Nepal Terai, remember their mythical heroes Dina-Bhadri as Mansukhdev, meaning 'God of humans' or 'Earthly God'. Earth and earthly environments host Dalit lives. Despite devaluation of material and human life, the earth's physicality and materiality embodies habitability, diversity, freedom, and change, offering unlimited promises for Dalits. They have been surviving on the earth's biosphere and natural resources, without threatening and subjugating other lives.

Labour and Mobility

Dalit writer Bama's autobiography narrates her entire life from childhood to the present – 'a life that is in touch with nature'[30] – growing up amongst people labouring from sunrise to sunset and surviving amidst hard and incessant labour. Bama journeys from a small village to the city, and from labouring in agricultural lands to being a high school teacher in Tamil Nadu. Manohar Mouli Biswas, a Dalit writer, recounts his days of growing up Dalit in a Bengal village, where his father, grandfather, and forefathers were agriculturalists, producing harvest in the 'wet earth', and feeding the state. He moves to Kolkata, works in different places/occupations, becomes the president of Bangla Dalit Sahitya Sanstha and editor of Dalit journals: 'Yet I remain in my familiar environment, in my familiar space – I want to remain here and so I write my autobiography.'[31] In Delhi's industrial areas, Dalit labourers migrated from Bihar, Uttar Pradesh, and Madhya Pradesh, worked in industries and acquired new occupational skills, continuously contesting the repressive nature of their work and environment through individual and collective efforts – from joining the trade union ranks to looking for alternative occupations and locations. Yet they also moved in and out of the industrial environment and labour unions, especially in the 2000s, when the city galvanized to unexpected environmental concerns and activism, and an emergent urban elite, drawn mainly from environmental, professional, and civil sectors, tried successfully to tie pollution problems with implicit caste and class factors.

These examples showcase the challenges in mapping the meanings of Dalit ecological spaces, and their representations in diverse regions, languages, traditions, and cultures of the country. Descriptions of nature abound in Dalit narratives of rain, river, fish, forest, hill, land, and agriculture. They celebrate the intrinsic beauty of nature and humankind's interaction with her, reflecting a unity between individual, society, and nature. However, all chapters in *Dalit Ecologies* illuminate the centrality of labour[32] – labouring experience; space and place; labouring bodies; interrelations between labour, industry, technology, and science; future of labour and work – in defining relations between society and nature, through the vantage point of caste, labour, and nature. Labour is experienced as a process of unending reproduction of social life by Dalit and low-caste people, which carries with it a whole economic and social process, and human production of the future. Caste labour is cruel. However, possibilities of labour, coupled with 'liberating' forces of technology, science, and development, are seen as undying flames of life through which natural and material worlds can be transformed. The present social aspect of labour proceeds along with its future aspect of a positive role of social and environmental consciousness. Dalit labour works and lives amidst both material and social relations with nature, an acute consciousness of caste processes and sources of domination, and a working out of new environmental ideas and objectives.

The circulatory out and in migration of men; an increase in women's participation in agriculture; transformed economic relations and public interventions; spread in education, knowledge, and social consciousness; and development of road, power, and telecommunication have made the world of Dalits mobile, effecting their relations with the world of nature. Purity-pollution has not been eliminated. A majority of Dalit labour is still tied with caste-based occupational hierarchy. But the haunting images of bonded labour on land, everyday violence over access to commons, and marginality in physical–social landscapes have been reduced. There is now a fertile ground for demands of technology and climate justice. Instruments of 'just' labour and justice issues in climate mitigation and adaptation are seen as important measures to improve human standings and society's relations with the surrounding nature, providing dignity to Dalit labour.

Casteization of Technology and Industry

Dalit Ecologies is situated at the crossroads of Dalit labour and the history of technology and industry in India. The conflict between perceiving technology and industry as a means of emancipation for Dalits and its historical production into a tool of caste-based subjugation emerges as a prominent theme across different regions and sectors, such as the stainless steel industry in the National Capital Region, tanneries in Uttar Pradesh, brick kiln industry in Haryana, and the urban development and waste production in Chandigarh's modernist landscape. Caste construction of technology is critical in several formal and informal sectors of economy.

Caste-based technology operates at multiple levels, perpetuating a hierarchical division among different castes. Dalits and people from lower castes are often confined to low-grade industrial skills, professions, and practices, while higher-paid positions such as engineers, supervisors, and designers are predominantly occupied by individuals from higher castes or classes. This caste-based division is also evident in the design of the industrial environment, which is typically controlled by powerful castes, with minimal participation from Dalits in shaping the workspace. In everyday working lives, both within and outside of industries, the possession and utilization of technology become indicators of one's position in society. Those who operate low-skilled machines and tools, whose physical bodies and labour are integral to the production system, and those who are excluded from high-end tasks find themselves excluded from technological innovation and progress. In fact, the world of technology rarely considers the work of Dalits as a significant concern.

For me, the tanners of Jajmau in Kanpur symbolize the complex and troubled relationship between Dalits and technology. On the one hand, tanners possess a rich history and expertise in the leather industry. On the other hand, the tannery's technological system, characterized by its sheer physicality and brutality, has inflicted numerous scars on their bodies. Despite being highly skilled, these tanners suffer from severe occupational diseases and fatalities. While they play a pivotal role in labouring and producing leather goods for both domestic and international markets, the detrimental aspects of their polluted nature and hazardous work environment have been largely ignored in the trajectory of industrial modernization in the country. The design, product development, safety measures, and training programmes in the industry have largely overlooked the issue of caste. As a result, the advancements made in these areas have failed to live up to the expectations and aspirations of Dalit labourers. The caste–industry–technology nexus continues to hinder the empowerment and upward mobility of Dalits in the leather sector.

The chapters shed light on the intricate historical dynamics and transformations within the realm of caste–technology relations. One noteworthy illustration of this can be found in the account of the leather industry, which encapsulates the colonial, post-colonial, and contemporary history of India. It chronicles the initial entry of Dalits into the informal industrial sector during the colonial era, and, subsequently, their significant involvement in both informal and formal sectors, which played a crucial role in the expansion of the burgeoning colonial–caste economy. In the post-colonial era, there was a remarkable surge in the caste–capitalist economy, with Dalit labourers participating in larger numbers and with greater intensity. Two major themes in the stainless steel and brick kiln industries are the high concentration of Dalit labourers, especially in the national capital and emerging industrial regions, and the advent of the new liberal and global economy in India. The expansion of these regions and industries presented opportunities for Dalit mobility and migration, but their technological systems were marked by caste-based segregation and exploitation. As labour rights and justice became increasingly demanded in certain regions and sectors, the stage was set for addressing issues of environmental and technological justice. Dalit social and support groups, along with segments of trade unions, emerged as agents for challenging technological injustices. In early independent India, Nek Chand crafted a compelling counter-narrative to the violence associated with modern industrialization. He accomplished this by expressing a subaltern perspective that embraced the earth, drawing upon memories, metaphors of labour, displacement, discards, and protests. Through his art, Nek Chand offered a powerful alternative vision that challenged the dominant industrial narrative and celebrated the resilience and creativity of the marginalized.

In both natural and social settings, Dalits have consistently employed technology, skills, and knowledge for their sustenance and progress. The advent of technology, the growth of industries, the rise of cities, and overall development have played crucial roles in enabling Dalits to break free from caste-based occupations and the confines of traditional society. However, the practice of industry and technology has also presented a dilemma for Dalits, as it is intertwined with caste relations. They face the challenge of navigating their lives while simultaneously harnessing the potential of technology for a profound transformation in their circumstances. The innovation and application of clean technologies within the social–industrial sphere, marked by the dichotomy of purity and pollution, have served as pivotal indicators of technological justice for Dalits and lower castes. Furthermore, the removal or replacement of existing traditional technologies with cleaner alternatives plays a crucial role in fostering technological progress. These factors serve as significant indicators of technological justice for Dalits, emphasizing the importance of ensuring their equal access to, acquisition of, and ability to apply clean technologies in order to address existing inequalities and promote fairness in the industrial and technological sphere.

Climate Justice and Dalitbahujan Anthropocenes

'Dalits are not afraid of climate change or any other natural disasters, because socially and economically, they have been leading a disastrous life forever. For them, disaster caused by upper caste Hindus is the real problem.'[33] This powerful statement of a Dalit activist connects with the critique of the Western notion of climate change, which primarily associates it with the fossil-fuelled Industrial Revolution and adheres to a teleological time and singularity within the Anthropocene narrative. *Dalit Ecologies* challenges this perspective by presenting counter-narratives from Dalits and low castes that delve into climate change and the Anthropocene through their historical context, lived experiences, and imagination.

In her memoirs, *The Weave of My Life*, Urmila Pawar, a renowned Dalit writer in Marathi, extensively portrays the harsh and ever-changing landscape and weather of the Konkan region on India's west coast during the mid- to late twentieth century. She vividly describes how the demanding journeys undertaken by Dalit women to sell their goods in local markets were deeply influenced by the weather and the landscape. Another perspective on Dalits and weather emerges from Vasant Moon's narrative of growing up in a *vasti*, an urban neighbourhood, which adds further dimensions to our comprehension of Dalits and their interactions with the complex slum culture. Gail Omvedt, a prominent anti-caste academician and activist who translated Moon's autobiography into English, expressed that she had never come across such a compelling depiction of the heat in Nagpur, despite her personal experiences with it.

The Dalit experience of climate predates the dominant history of climate change, which could even be dated back to famines, deaths, and disasters in the nineteenth century. In addition to the historical context, which has strong roots in caste and colonialism, Dalits perceive climate through the lens of the everyday and the ordinary. They embrace a temporality that is not teleological, that is, it does not follow a climate science-determined linear progression. Their language to express their relationship with climate is one of caste affect and incomprehension, capturing the emotional and experiential aspects rather than solely focusing on understanding and adapting to it. In the globalized and liberalized economy of present India, the situation of Dalit labourers in the brick kiln industry in Haryana serves as a significant case study. They are among the most severely impacted population in the region due to the increasing emissions from the industry and the associated heat. In these contexts, Dalit organizations are increasingly advocating for the recognition of caste as a critical category within the discourse on climate change at national and international levels. This book endeavours to explore the political stance of Dalit organizations regarding climate change, which differs from the perspectives of mainstream environmental and civil society groups, as well as climate experts. The slogans 'Dalit Climate Justice' and 'Caste Justice is Climate Justice' have gained prominence as rallying cries, frequently voiced by Dalits to assert their presence and perspective within climate action activities. These slogans encapsulate their demand for equitable treatment in addressing climate issues, highlighting the inseparable link between caste-based justice and environmental justice.

The concept of the Anthropocene has faced critical scrutiny from numerous commentators, despite recognizing its importance. These commentators have raised concerns about its tendency to universalize and overlook the diverse interests and conflicting worldviews – 'differentiated people with contradictory interest, opposing cosmos' – among diverse groups of people. They have also highlighted the 'differential responsibilities' for climate change.[34] In this regard, Heather Davis and Etienne Turpin have observed that 'the Anthropocene is not simply the result of activities undertaken by the species *Homo sapiens*, instead, these effects derive from a particular nexus of epistemic, technological, social, and political economic coalescences figured in the contemporary reality of petrocapitalism'.[35] There have been calls for an indigenization and provincialization of the Anthropocene concept. This entails dismantling the European, white supremacist, and heteropatriarchal frameworks and discussions surrounding Earth systems. The aim is to prioritize and include the perspectives, knowledge, and experiences of indigenous communities and localized contexts in understanding and addressing the challenges of the Anthropocene. By decentring dominant narratives, this approach seeks to foster more inclusive and equitable dialogues about our relationship with the planet.[36]

Dalit Ecologies exemplifies how Dalit and low-caste individuals can become crucial collectors and narrators of the Anthropocene and climate change. The making of Nek Chand and the Rock Garden in the modern city of Chandigarh illustrates the intricate and interconnected processes of region–scale dynamics; historical–national factors; and the caste-, capital-, and state-driven acceleration of human activities. These processes have led to extensive displacement, generated massive amounts of waste, and established a city that radically changed the ecological interconnectedness of life.

Nek Chand, through his identification with the marginalized, discarded, and wasted humans and nonhuman beings, as well as rocks, mountains, and rivers, was able to envision a creative and critical alternative to the dominant narrative of the Anthropocene. His artistic expression and imagination opened up possibilities for reimagining our relationship with the environment and offered inclusive approaches to our interactions with the natural world. It is important to note that, similar to many Dalit writers and cultural activists, Nek Chand utilized various artistic mediums such as art, architecture, sculptures, and images to create a space for narrating a counter-history of the Anthropocene. This narrative was characterized by its relational and intersectional nature, addressing the concerns of diverse segments of Chandigarh society. Thus, the people of Chandigarh became mobilized on multiple occasions to prevent the demolition of the Rock Garden. The Garden became a symbol of resistance and a source of inspiration for the community. It underscores the power of Dalit art and culture in mobilizing communities, generating awareness, and instigating meaningful dialogue about *Dalit Ecologies*.

Eco-literary Traditions

The Dalit and low-caste movements have achieved remarkable progress in empowering marginalized peoples, providing them with intellectual and cultural resources, and instilling the confidence to challenge dominant forms of Indian environmentalism. These movements, along with the contributions of anti-caste activists and writers spanning from the Phule–Periyar–Ambedkar era to the present century, have examined various aspects of natural resources, villages, communities, agriculture, industry, labour, technology, science, and hierarchies within both the natural and social order.

Specially since the 1990s, there has been a surge in Dalit literary works across India in various languages. The genres of Dalit literature, including fiction, poetry, self-narratives, autobiographies, visual art, plays, songs, ballads, films, and folklore, have flourished. Education, political activism, publishing platforms, and advancements in technology have played significant roles in nurturing and showcasing this abundant talent. Numerous Dalit writers, encompassing school and college teachers, journalists, students, activists, factory workers, peasants, agricultural labourers, nuns, and fieldworkers, have transformed the literary landscape through their creative expressions. In addition to the presence of independent Dalit publishers, there has been a growing interest from English and other commercial publications in promoting Dalit writings. This indicates a widening recognition of the significance and impact of Dalit voices in contemporary Indian literature.

Dalit Ecologies delves into a diverse wealth of folk traditions, regional and linguistic expressions, and creative literature to unravel how Dalit and low-caste writers, often situated outside the domain of environmental publications, have utilized their experiences, imagination, and language to construct their perceptions of environmentalism. These writers also employ these tools to fashion alternative metaphors pertaining to rural, urban, industrial, and spatial contexts, which serve as fertile ground for the creation of Dalit environmental culture and politics. By doing so, they build new pathways for advocacy, campaign, and mobilization, fostering a sense of agency, empowerment, and resistance.

This book features several chapters that draw extensively from the life narratives and other writings of prominent Dalit and low-caste writers such as Bama, Manohar Mouli Biswas, Siddalingaiah, Adwaita Mallabarman, Nek Chand, Urmila Pawar, and Vasant Moon. While these writers represent only a fraction of the vast number of Dalit and low-caste writers across the country, their inclusion in the book provides valuable insights into the intricate relationship between nature, culture, and society, and its impact on environmental politics. Although this sample size of writers is limited, their works shed light on the intricate interplay between human experiences, the environment, and social structures.

Indeed, each Dalit autobiography encompasses a wide array of personal and theoretical approaches to meticulously unravel the intricate connections between nature and caste. To exemplify this, let us explore the autobiography of Dalit writer Sharankumar Limbale,[37] shedding light on four key dimensions – the depth and breadth of Dalits' intimate encounters with nature and their celebration of it through rich and colourful expressions; the overarching presence and the painful experience of caste in nature; the central role of labour in understanding Dalits' relationship to nature in their everyday lives; and the conjunctions between animal, animality, and Dalit identity formation. Beautiful descriptions of nature abound in Limbale's narrations of rain, river, and fish: 'The sky too fell into the water. I was transformed into water … Wave after wave rose on the surface of the water. Pathway after pathway opened in my heart.'[38]

At the same time, Limbale reflects on this interrelationship between nature and caste, power and place:

> Play over, we settled down to eat. Boys and girls from the high castes like Wani, Brahmin, Marwari, Muslim, Maratha, Teli, fishermen, goldsmiths and all the teachers, about hundred or so sat in a circle under a banyan tree. We, the Mahar boys and girls, were asked to sit under another tree....
>
> Even the tree we sat under was tattered like us. Whenever the wind shook its branches it produced waves of hot air that hit our faces. We sat in its broken shadows.[39]

Limbale describes his labour in and around the river:

> We used to go to the river to catch fish. Sometimes we caught fish in a sari, at other times in a bowl and vessel. Sometimes we made puddles in the sand, then removed the water and caught the fish. We were bored of angling of fish. At night we cut the fish in the light of a bonfire made with discarded tyres. During summer when the river shrank, there were small puddles instead of flowing water. We then stuffed the puddles with branches and leaves of the sher plant. The sap of the sher plant is poisonous so with it the water in the puddles too became poisoned. One by one the fish died, wriggling before they fell quiet. Then they floated to the surface. The dead fish sparkled like silver. Our hands stank from gathering the dead fish.[40]

Limbale perceives his community through an analogy to animals, 'which had lived the life of cats and dogs for thousands of years'.[41] Limbale comments, 'Hindus see the cow as their mother. A human mother is cremated but when a cow dies they need a Mahar to dispose it of.'[42] After giving a long account of pulling the carcass of a buffalo calf, at another place he bitterly says, 'What sort of a life was this? Drag dead animals, skin them, and then eat their flesh.'[43] The disposal of dead animals has an everyday, complex presence in Dalit lives. It signifies a forced, polluted, and traditional occupation, a source of food, meat, and income, and a symbol of death, decay, blood, and smell.

There are numerous examples that illustrate the point being made. However, the main idea is that *Dalit Ecologies* aims to compile a comprehensive collection of regional, literary, and language writings by Dalits, in order to represent their historical and contemporary environmental perspectives. In India, environmental academics have traditionally relied on the authority and authenticity of extensive archival and field research. However, Dalits have raised concerns about the omissions and exclusions resulting from these research methods, as well as the specific production and institutionalization of knowledge. Dalits create their own archives, consisting of autobiographies and literature, which provide greater insight into their natural and material lives and a wide range of experiences. This opens up new avenues for studying Dalits, caste, and ecology.

Relying on Dalit writings and extensive fieldwork, *Dalit Ecologies* attempts to study the relationship between Dalits and low castes and their physical environment, and its manifestation in their everyday life, in concerns and struggles for environment justice and recognition on the one hand, and their continued experiences of being the environmental 'other' in casteized nature on the other. In Chapter 2, I deal with India 'nature writing', which has traditionally encompassed ecology, geography, and sacrality, and has often missed Dalit literary traditions. In the last few decades, environment literature has expanded its horizons to consider intersections between ecology, society, and culture. However, the question still remains: Why is there no recognition of ecological underpinnings in the writings of subordinate castes by the wider canon of environmental literary sphere? I take Dalit autobiographies from different regions and languages to highlight an unexplored aspect of Dalit writings, thus widening the scope and perspective of environment literature and providing a distinct perspective from the margins. Through the lens of eco-literature, eco-criticism, and eco-justice, I underline how nature's beauty and caste burden; space and identity; land and bondage; and social injustice and environmental 'othering' are significant features of these life narratives. I weave certain select themes like nature's beauty, caste exploitation, labour, and animals to explore the pain and stigmatization, along with a vibrancy and dynamism, in Dalit eco-narrations of the self.

Chapter 3 closely studies Dalit folktales, which are an important part of Indian folkloristic categories. They have existed since long, often in a complex relationship with the dominant culture. The development of Dalit cultural, political, and public spheres in the recent past has galvanized a dynamic Dalit system of organization of folktales in performances, festivals, and protests, leading to a diffusion of its contents and forms, and intimate connections with the everyday life of the community. The folktale has also increasingly become an expression of Dalits' environmental risks, conflicts, and rights. Its mythical characters have been transformed into symbols of ecological ancestors with immense physical, natural, and spiritual skills, who have the courage to liberate the community from oppressive ecological and social systems. Drawing on the folktale of two Musahar brothers, Dina and Bhadri, widely circulated in the

Indo-Gangetic Plains of India and Nepal, I shed light on how folklore has become a way for Dalits to assert their environmental rights through their own motifs, everyday memories, and ecological ancestors. The ecological symbolism of Dalit folklore is dynamic and creative.

Chapter 4 is a close reading of a literary text, *A River Called Titash*, and unravels some layers of commons and rivers in the making of Dalit environments. Written by Adwaita Mallabarman, a poor Malo Dalit, and autobiographic in a wider sense, the novel weaves a complex narrative of nature, place, commons, and community at the turn of the twentieth century. Placing archives of Dalits and eco-criticism, novel and river, caste and commons in tandem with each other, it shows how Dalit environments comprise a unique set of analytic, where earth, commons, culture, and place-attachment-and-movement strings build their traction towards nature.

Chapter 5 turns to discussions on the Anthropocene, in which 'untouchables' and lower castes have never been included. Anthropocene can be critiqued, as well as expanded, by including the experiences and imaginations of people living on the peripheries and discriminated on the basis of their birth and occupation. I use the concept of Dalitbahujan Anthropocenes, which creates a base to understand the complex interplay between the post-colonial state, modernity, nature, and marginal people. Inspired by anti-caste and anti-race thinking and initiatives, I expand the existing discourse on Anthropocene. Dalits and low-caste people have been central to transformations in ecologies and human communities, constituting what Dalit thinker and freedom fighter B. R. Ambedkar termed the 'primitive accumulation in the modern caste economy'. Caste economy underscores the 'untouchable' and 'polluted' transformations of the earth's geological and ecological processes, in relation to the accumulationist prerogatives of upper-caste/class people in the modern age. Dalitbahujan Anthropocene is also a means of thinking about the radical experiences, critical interventions, and practices born of Dalit mobility, migration, and mitigation, in relation to matters of geographic and environmental vulnerability. This vision of Dalitbahujan ecologies as sites of possibility for alternative relationships holds bearing on future visions of robust sustainability, reciprocity, Dalitness and (non) humanness, Dalit–Bahujan solidarity, and environmental justice. Through Nek Chand and his creation of the Rock Garden in Chandigarh city, I see how displacement, violence, waste, rock, and human creativity can be an axis to anchor experiences of certain castes and communities in the modern industrial age, which also pave the path for radical imaginations of future life on the earth.

Chapter 6 is about the industries and industrial areas in the national capital, Delhi. The industries have developed in a particular form of exploitation of human, natural, and social structures, where nature is industrialized and caste is naturalized. The division of labour in a factory interacts with the caste hierarchy in society, where Dalit working body, physicality, and social position play an important role within the production process, under the control of capital. The environmental politics against pollution can create an industrial and social environment within which the polluting industries and 'impurity' of Dalit labourers' bodies become synonymous with each other. Taking the analytic frameworks of Karl Marx and Ambedkar, I explain how Dalit labourers in stainless steel utensil factories move in and out of the industrial ecosystem and how they contest the repressive nature of their work and environment.

Chapter 7 deals with issues of technology, industry, caste, and Dalits through a case study of tanneries and tanners in Uttar Pradesh. In environment justice, technology and science have acquired a particular importance in the past two decades. Dalits and low castes, who are tied with birth-based 'dirty', 'polluted' occupations, have a great urge for technology justice, in order to free them from their degraded positions. In many industries, technology designs are caste-shaped, and through these designs industries also contribute to ordering society. Tanneries and tanners today are a testimony of the fact that caste power relations continue to be alive through a reshaping and reproducing of technology, forms of industrial organizations, and social arrangements. Dalit labour emphasizes the need to change their lower labour and social positions by initiating radical processes of de-Dalitizing and de-casteizing the tannery technological system.

Chapter 8 moves to climate change and climate justice. Intersections between weather, climate, and caste in India have initiated a dialogue on the need to recognize caste as an important category, similar to gender or race, in contemporary climate change discussions and activities. Climate scientists and experts tend to emphasize the universality and social generality of climate change, and in the process often have a skewed social understanding of the caste question, as they either take a supposedly 'caste-neutral' approach or work with caste stereotypes, thus implicitly revealing a 'climate casteism'. A fieldwork-based case study of brick kilns in Haryana demonstrates how economy, industry, labour, and occupation-related climate hazards are intimately interwoven with the 'caste economy', which also calls for an analysis of the *savarna* effect on economic decision-making pertaining to climate change. Through interactions with a leading Dalit organization in India, the attempt is to understand how Dalits would like to articulate climate change politics, based on their knowledge, research, and campaigns.

Chapter 9 is an inquiry on caste and race, Dalits and blacks, and the common ground between them in the environmental field, which has so far been neglected. Drawing from a plural framework of environmental justice and histories of environmental struggles amongst the African Americans, it focuses on historical and contemporary ecological struggles of Dalits. It contemplates on how their initial articulations under the rubric of civil rights developed into significant struggles over issues of Dalit access, ownership, rights, and partnership in natural resources, where themes of environmental and social justice appeared on the forefront. The intersections between Dalit and black ecologies, the rich legacies of Black Panthers and Dalit Panthers, and their overlaps in environmental struggles open for us a new historical archive, where Dalit and black power can talk to each other in the environmental present.

In concluding my introduction, I feel compelled to briefly examine my personal identity, journey, and positionality within the context of writing about *Dalit Ecologies*. I am a forward-caste Bhumihar Brahmin by birth. In the mid-1980s, I embarked on a career in professional journalism with the *Navbharat Times*, Delhi, covering environmental and labour issues. People, poverty, class, and environmental struggles were my main rubrics. During this time, the writings of scholars such as Ramachandra Guha, Darryl D'Monte, Anil Agarwal, and Praful Bidwai significantly influenced my understanding of critical theoretical, political, and historical frameworks pertaining to environmental research and writing. Their perspectives, rooted in environmentalism and social movements, provided me with invaluable guidance.

However, it is crucial to acknowledge a critical gap that existed in my approach, as well as that of many other journalists and researchers. While we scrutinized dominant paradigms of development and the state, I failed to fully recognize the significance of the politics of representation, as it pertained to writers and journalists from dominant castes. Looking back, I now realize that not only did I adopt a political lens focusing on class and poor but I also lacked awareness regarding the potential biases and limitations associated with representation by individuals from privileged castes. This oversight prevented me from fully understanding the experiences and perspectives of marginalized communities, including Dalits, within the environmental discourse. Recognizing this shortcoming is essential, as it highlights the need for critical self-reflection and a commitment to rectifying the imbalances that can inadvertently perpetuate inequality.

In the 1990s, some critical shifts occurred, which slowly came to affect writing and research on caste. First, with the meteoric rise of Dalit politics and assertion of Dalit voices, particularly in north India, the question of caste became important in the academic sphere. Second, with the Mandal Commission,[44] a new caste discourse emerged in the political–public arena. Yet most of these, too, missed the environment dimension. The struggles of Bahujans and Dalits were predominantly portrayed as political and social activities. Further, the writings focused on the political–public sphere and cataclysmic events, as opposed to the ordinary, the mundane, everyday lived lives of Dalits. Equally, they often left unexamined Hindu upper-caste position and privilege, designed as they were around environmental protests and struggles. Caste was seen as a synonym for Dalit, as a burden and concern of Dalits alone, which did not impinge on the lives and histories of dominant castes.

During my extensive fieldwork and research on environmental issues, particularly my studies on figures like Anna Hazare and Ralegan Siddhi, as well as my research on the anti-Tehri and anti-Sardar Sarovar dam movements during the 1990s and 2000s, I became acutely aware of the challenges in making universal claims about environmental politics. I recognized the critical role of caste power and the marginalization of Dalits, whose concerns often took a backseat to those of dominant castes. In an effort to address these issues, I consciously adopted a caste lens, focusing on the everyday experiences, representations, and agency of Dalits. This shift in perspective was influenced by my deep and enduring friendships with Dalit activists such as Ashok Bharti and Ranji Tilak, as well as our collective involvement in the National Conference of Dalit Organizations (NACDOR). Together, Ashok Bharti and I initiated the World Dignity Forum through NACDOR, collaborating with various Dalit organizations in India and South Asia. Additionally, I occasionally partnered with the National Campaign on Dalit Human Rights (NCDHR) at the Asia Social Forum, which further expanded avenues for cross-caste solidarity. Building cross-caste solidarities in both my personal and professional lives is an ongoing and arduous journey, filled with challenges and constant change.

At the time of heightened Hindutva mobilization, caste-based atrocities, and a dynamic Dalit–low castes political and social landscape, not to mention the globalization of anti-caste movements and myriad activism around environmental and climate justice, *Dalit Ecologies* tries to present arguments and case studies that highlight the Dalit mode of thinking and representation on environmental issues and shows how Dalits and low-caste lives matter in the struggles for a just and sustainable future.

Notes

1. In 2004, for the first time, I delved into this subject as I presented a seminar on 'Where Are Dalits in Indian Environmentalism?'. Thereafter, in 2012, I published a research article 'Dalits and Indian Environmental Politics', which helped me write the book *Caste and Nature*, published in 2017. I reworked this book in Hindi, which was published in 2020. Several research papers, talks, panel discussions, and interviews of mine have appeared on the subject since 2015. For details, see 'National Seminar on Dalit Studies and Higher Education: Exploring Content Material for a New Discipline', Deshkal Society, Delhi, 28 February 2004; Mukul Sharma, 'Dalits and Indian Environmental Politics', *Economic and Political Weekly* 47, no. 23 (2012): 46–52; Mukul Sharma, *Caste and Nature: Dalits and Indian Environmental Politics* (New Delhi: Oxford University Press, 2017); Mukul Sharma, *Dalit aur Prakriti: Jati aur Paryavaran Aandolan* (New Delhi: Vani Prakashan, 2020).

2. I introduced this term in my three-part webinar 'Dalit Ecologies Series', that is, 'Seeing through History', 'Towards an Anthology of Folklore', and 'Future of Dalit Ecologies' in 2020–21 under the 'Anti-Caste Politics and Environmental Justice' initiative. For details, see Seshadripuram Evening Degree College, Bengaluru, https://www.youtube.com/channel/UCwLDfNwoFVV8VSFt80mXghg/videos-, accessed on 6 January 2024.

3. This was a six-chapter web initiative, from September 2020 to February 2021, in which fifteen Dalit and anti-caste intellectuals, activists, and academicians spoke on different aspects of caste and environmental justice, and altogether 1,500 people participated in the series. For details, see Seshadripuram Evening Degree College, Bengaluru, https://www.youtube.com/channel/UCwLDfNwoFVV8VSFt80mXghg/videos-, accessed on 6 January 2024.

4. The literature on these themes is considerable. For some important studies, see Nathan Hare, 'Black Ecology', *Black Scholar* 1, no. 6 (1970): 2–8; Leilani Nishime and Kim D. Hester Williams, eds., *Racial Ecologies* (Seattle: University of Washington Press, 2018); Lara Stevens, Peta Tait, and Denise Varney, eds., *Feminist Ecologies: Changing Environments in the Anthropocene* (Switzerland: Palgrave Macmillan, 2018).

5. In 2022, *Environment and Society: Advances in Research* brought out a special issue on 'Global Black Ecologies' to render visible the work of black communities in creating alternatives in environmentalism. The journal issue was part of a collaborative effort, which included the Black Ecologies series, the Black Ecologies Initiative, *Making Livable Worlds*, and

a zine publication. J. T. Roane, an editor of the issue and the author of 'Plotting the Black Commons', introduced me to this special issue.

6. Achille Mbembe, *Critique of Black Reason*, trans. Laurent Dubois (Durham, NC: Duke University Press, 2017).

7. I am a native of Bihar, living in Delhi and working in Haryana. Fieldwork in Bihar and Delhi has been an ongoing work since 2000. Extensive fieldwork in Uttar Pradesh, Haryana, and Punjab took place between 2015 and 2022.

8. I identify at least five such tension points: apologist and recuperative Brahminism, and a stream of modern environmentalism in modern India; closing eyes to Dalit environmental voices and visions; seeing Dalits as cheerleaders of development and modernity, detached from environmental discourses; glorifying and 'humanising' degraded occupations in which Dalits are the main participants; and mainstream environmentalism being seen as elitist by Dalit thinkers, devoid of their concerns. For details, see Sharma, *Caste and Nature*.

9. Dalits and anti-caste thinkers have been articulating this view in the recent past, in the wake of emergence of conservation, urban, and climate change environmentalism. For details, see Gopal Guru, 'Moral Significance of Justice: Foregrounding Environmentalism in India', Public Lecture at Nehru Memorial Museum and Library, New Delhi, 26 September 2013.

10. The term 'hybrid nature' is used by Kiran Asher in her outstanding research on Afro-Colombians of the Pacific Lowlands. According to her, black social and environmental movements champion causes of biodiversity conservation and sustainable resource management in tropical rainforest by adopting a hybrid of cultural politics, mass mobilization, and latest technocratic development intervention. Quoting the works of Escobar on black struggles in the Colombian Pacific, she explains how the process of black cultural organizing, state policies, and development interventions open up new spaces for environmental struggles. For details, see Kiran Asher, *Black and Green: Afro-Colombians, Development, and Nature in the Pacific Lowlands* (Durham: Duke University Press, 2009).

11. I am grateful to environmental historian Dianne D. Glave, whose research on the African American environmental heritage has opened a new world to me. For one important example of her work, see Dianne D. Glave, *Rooted in the Earth: Reclaiming the African American Environmental Heritage* (Chicago, Illinois: Lawrence Hill Books, 2010). I have learnt immensely from the writings of and interactions with J. T. Roane, Dorceta E. Taylor, Robert D. Bullard, and Kimberly N. Ruffin.

12. David Schlosberg, *Defining Environmental Justice: Theories, Movements, and Nature* (Oxford: Oxford University Press, 2007), XIII.

13. For more discussion on this, see Nancy Fraser, *Justice Interruptus: Critical Reflections on the 'Postsocialist' Condition* (New York: Routledge, 1997); and Axel Honneth, *The Struggle for Recognition: The Moral Grammar of Social Conflicts* (Cambridge: MIT Press, 1995).

14. Robert M. Figueroa, 'Bivalent Environmental Justice and the Culture of Poverty', *Rutgers University Journal of Law and Urban Policy* 1, no. 1 (2003), 27–44.

15. Schlosberg, *Defining Environmental Justice*, 66.

16. Robert D. Bullard, *Dumping in Dixie: Race, Class, and Environmental Quality* (Boulder, CO: Westview Press, 2000).

17. For a fine articulation and elaboration of this subject, see Ryan Holifield, Michael Porter, and Gordon Walker, eds., *Spaces of Environmental Justice* (West Sussex: Wiley-Blackwell, 2010).

18. There were several significant developments. For example, First National People of Color Environmental Leadership Summit in 1991; Second People of Color Environmental Leadership Summit in 2002 that adopted the 'Principles of Working Together'; and focus on the 'Jemez Principles for Democratic Organizing'. For details, see Christopher W. Wells, ed., *Environmental Justice in Postwar America: A Documentary Reader* (Seattle: University of Washington Press, 2018).

19. There is extensive writing on environment justice in Latin America. See, for example, J. Timmons Roberts and Nikki Demetria Thanos, *Trouble in Paradise: Globalization and Environmental Crisis in Latin America* (New York: Routledge, 2003).

20. David V. Carruthers, ed., *Environmental Justice in Latin America: Problems, Promise, and Practice* (Cambridge: MIT Press, 2008).

21. Juanita Sundberg, 'Tracing Race: Mapping Environmental Formations in Environmental Justice Research in Latin America', in *Environmental Justice in Latin America*, ed. David V. Carruthers (Cambridge: MIT Press, 2008), 25–48.

22. Melissa Leach and Robin Mearns, eds., *The Lie of the Land: Challenging Received Wisdom on the African Environment* (Portsmouth: Heinemann, 1996), 19–20.

23. Jacklyn Cock and David Fig, 'The Impact of Globalisation on Environmental Politics in South Africa, 1990–2002', *African Sociological Review* 5, no. 2 (2021), 15–35.

24. David A. McDonald, 'Environmental Racism and Neoliberal Disorder in South Africa', in *The Quest for Environmental Justice: Human Rights and the Politics of Pollution*, ed. Robert D. Bullard (San Francisco: Sierra Club Books, 2005), 255–78.

25. Darren Ranco and Dean Suagee, 'Tribal Sovereignty and the Problem of Difference in Environmental Regulation: Observations on "Measured Separatism" in Indian Country', *Antipode* 39, no. 4 (2007): 691–707.

26. I focus on Nek Chand and Rock Garden in a separate chapter.

27. Thomas Berry, *The Dream of the Earth* (Oakland: Sierra Club Books, 2016), 42.

28. Carl Anthony, 'Reflections on the Purposes and Meanings of African American Environmental History', in *'To Love the Wind and the Rain': African Americans and Environmental History*, ed. Diana D. Glave and Mark Stoll (Pittsburgh: University of Pittsburgh Press, 2006), 200–09.

29. Earth worship is found widely amongst Dalits and low-castes in different regions and traditions. This has also been documented in research. For example, James Massey, an author and well-known Dalit Christian theologian, gives an example of his village Zafferwal (Punjab) where Dalits worship a man-made dome-shaped mound of earth called Bala Shah. This mound faces eastwards and is placed under a tree or an open spot with an earthen lamp burning over it and also a red flag on top of it. On an auspicious day, the devotees worship here with their offerings. The burning lamp and the red flag are for security from all kinds of evil forces. For details, see James Massey, *Dalit Theology: History, Context, Text and Whole Salvation* (New Delhi: Manohar Publications, 2014), 152. In an autobiographical novel on Dashrath Manjhi, an icon of Dalits in Bihar, we find several references of Earth celebrations. For details, see Nilay Upadhyay, *Pahar* (New Delhi: Radhakrishan Prakashan, 2015). Also, Rahul Ghai, Arvind K. Mishra, and Sanjay Kumar, eds., *The Marginalized Self: Tales of Resistance of a Community* (Delhi, Primus Books, 2020).

30. Bama, *Karukku* (New Delhi: Oxford University Press, 2012), 138.

31. Manohar Mouli Biswas, *Surviving in My World: Growing Up Dalit in Bengal* (Kolkata: Samya, 2011), xxi.

32. On the question of labour and its significance for ecological/environmental analysis, see William Cronon, ed., *Uncommon Ground: Towards Reinventing Nature* (New York: W. W. Norton & Company, 1995); Kimberly N. Ruffin, *Black on Earth: African American Ecoliterary Traditions* (Athens: The University of Georgia Press, 2010); Donald S. Moore, Jake Kosek, and Anand Pandian, eds., *Race, Nature, and the Politics of Difference* (Durham: Duke University Press, 2003).

33. I have dealt on this issue in Chapter 8.

34. Bruno Latour, 'The Anthropocene and the Destruction of the Image of the Globe', Gifford Lectures on Natural Religion, University of Edinburgh,

18–28 February 2013, https://www.ed.ac.uk/arts-humanities-soc-sci/news-events/lectures/gifford-lectures/archive/series-2012-2013/bruno-latour, accessed on 6 January 2024.

35. Heather Davis and Etienne Turpin, 'Art and Death: Lives between the Fifth Assessment and the Sixth Extinction', in *Art in the Anthropocene: Encounters among Aesthetics, Politics, Environments and Epistemologies*, ed. Heather Davis and Etienne Turpin (London: Open Humanities Press, 2015), 7.

36. For details, see Kathleen D. Morrison, 'Provincializing the Anthropocene', *Seminar*, https://www.india-seminar.com/2015/673/673_kathleen_morrison.htm, accessed on 6 January 2024; Zoe Todd, 'Indigenizing the Anthropocene', in David and Turpin, *Art in the Anthropocene*, 246.

37. Sharankumar Limbale, *The Outcaste: Akkarmashi* (New Delhi: Oxford University Press, 2003).

38. Ibid., 26–27.

39. Ibid., 2.

40. Ibid., 67–68.

41. Ibid., 103.

42. Ibid., 14.

43. Ibid., 53.

44. The Mandal Commission granted 27 percent reservations to 'Other Backward Classes' (OBCs) in all government jobs, thus making reservations for Dalits, Tribals, and OBCs a total of 49 per cent.

2

Literature

'My World Is a Different World'

Caste and Dalit Eco-Literary Traditions

Dalit and low-castes literature vividly portrays their ecological experiences, perspectives, and aspirations through a wide range of forms and themes. Within their literary works, there is a rich tapestry of both human and non-human elements, encompassing figurative and imaginative landscapes, as well as representations of natural and social environments. They reveal notions of belonging and marginalization in relation to nature, examining issues such as access and alienation from natural resources. The discourse on caste and environmental justice has made efforts to acknowledge and appreciate the Dalit eco-literary tradition. This involves critiquing the prevailing nature writing traditions and analysing Dalit literature through ecological lenses. Unlike the pursuit of an idyllic era of environmental harmony, this literary tradition aims to carve out Dalits' distinct language, memory, and vibrancy.

In her autobiography, Bama, the first published Indian Tamil Dalit woman writer,[1] describes herself as Karukku, meaning palmyra leaves with their serrated edges on both sides, like a double-edged sword challenging its oppressors. Her life of cruel caste oppression within the Catholic Church was like that of 'a bird whose wings had been clipped' and her recovery from social and institutional betrayal felt 'like a falcon that treads the air, high in the skies'. Manohar Mouli Biswas, a Bengali Dalit writer from West Bengal, writing for over three decades, thought of himself as a water hyacinth.[2] Initially he had named his autobiography *Prisnika* (water hyacinth). Later he renamed it as *Life and Death of Prisnika*, while expressing this self-identity as deeply hurtful for him. The world of Dr Siddalingaiah, one of India's foremost Dalit writers in Karnataka, breathes and dies in *ooru keri* – a separate space in a village or a city, where a Dalit resides – so much so that his autobiography is titled after this place.[3]

Growing up as a 'Namashudhra'[4] in a peasant family in West Bengal, a Dalit woman in Tamil Nadu, a Dalit man in southern Karnataka – a diverse range of contemporary Dalit writers, working in different regions and languages, identify themselves in nuanced ways with nature in their literary works. Their rich and varied stories – of discrimination and creation, humiliation and heartbreak, hope and freedom – are also nature stories, where they forge a relationship with the environment in diverse ways. In the process, they navigate natural, social, personal, and spiritual streams of consciousness that are markedly different from the mainstream. Even when not overtly visible, they move from one destination to another, connecting their emotions with nature. However, Dalit eco-literary traditions have been marginalized and overlooked by Indian environmental writing.

Based on three autobiographies by prominent Dalit authors from different backgrounds, regions and languages – Bama, Manohar Mouli Biswas, and Siddalingaiah – I will look at the images, vocabularies, metaphors, memories, and meanings associated with Dalit experiences of nature. Tracing nature's imagery, I will locate how and why certain principal elements of environment – land, water, forest, animal, and space – appear in significant shades in Dalit self-narratives. Some distinct features of Dalit eco-literary traditions are examined here and emphasized so that the blind spot of excluding Dalit writings from Indian eco-literary genres can be redressed by exploring the relationship between caste, nature, Dalits, and environmental imagination.[5] The emergence of Dalit literature can be seen as intersecting with eco-literature, eco-criticism, and eco-justice, and this highlights the ways in which Dalits are also imaginative ecological writers. The autobiographies chosen are selective, but they are representative of their region, language, and politics.

Dalit Depository: Literature of Their Own

The upsurge in Dalit literary writings in Indian languages across the country, especially since the 1990s, has been widely acknowledged. Dalit fiction, poetry, self-narration, autobiography, visual art, play, song, ballad, film, and folklore have blossomed, with education, political assertion, publication, and technology playing a significant role in this efflorescence of talent. Many Dalit writers – school and college teachers, journalists, students, activists, factory workers, peasants, agricultural labourers, nuns, and fieldworkers – have changed the literary landscape through their creative work.[6] Besides independent Dalit publishers, there has been a growing interest of English and other commercial publications in Dalit writings. This surge has not been sudden; it contributes to a larger history of Dalit literary writings in India.[7]

Dalit literary works have been characterized eclectically – as personal experiences and self-reflections; as narrations of pain, humiliation, subjugation, and anger; as concerns for civil rights; as discourses of social justice; as myths, memories, and fantasies; as means of resistance and challenge; as literature of the masses; as affirmations of identity; and as cultural, religious, and social movements from below to build a new future.[8] The question of what constitutes Dalit literature and who is a Dalit writer has also been debated extensively.[9] Dalit writing is understood as a contribution by members of the Dalit community, bearing witness to their experience of deprivation. It has also been said that Dalit 'writing is about Dalits by Dalit writers with a Dalit consciousness'.[10] The difference between Dalits and non-Dalits writing on Dalit issues has been sharply parsed as 'the difference between a mother's love and a wet nurse's love'.[11] There has been no definite or single way to define Dalit literature.

Dalit life-writing – autobiography, memoir, testimony, collective biography – has been a central genre of Dalit literature. The autobiographical form is present in a large portion of Dalit literature because of the 'confessional characteristic' often adopted by Dalit authors. Dalit life-writings have received critical acclaim due to their unique literary, social, political, and cultural richness; they have been widely translated and appreciated for providing deep insights into the life journeys of Dalits and their search for a dignified place in society. Raj Kumar states that Dalit personal narratives often 'capture the tensions which grow out of a continuous battle between "loss of identity" and "asserting of self"'.[12] Debjani Ganguly argues that they 'herald the emergence of Dalit personhood'[13] through a realization of full citizenship. M. S. S. Pandian underlined that they 'underscore the self-conscious ordinariness of the lives narrated', while bringing into focus how certain lives have been treated as unworthy and trivial by dominant practices of social sciences. They also frequently erase 'specificities of places and events' and mask 'them with a veil of anonymity' so that they can be 'anywhere and anytime'.[14] Sharmila Rege made a persuasive case for the significance of Dalit women's testimonies in challenging the prevailing singular communitarian notion of Dalit community.[15]

These widely informed and rich assessments of Dalit life-writings cover diverse aspects; yet they do not frontally acknowledge or recognize the ubiquitous presence of nature, ecological symbols, and environmental thought in these narratives. Several elements and images revolving around nature have been noticed by many perceptive reviewers of this literature, but there is no concerted reflection of its wider implications for environmental histories and scholarship. Before delving into the ecological richness of Dalit life-writings, let me briefly describe the state of dominant environmental literary writings in India and how they have dealt with Dalit literature.

Nature Narratives: Mapping Indian Environmental Literary Sphere

It is challenging to map representations and meanings of Indian environmental literary space in diverse regions, languages, traditions, and cultures in the country. Even though the sea of literature has aesthetically engaged with nature, a study of the interrelationship between literature and environment is quiescent in India. However, there are broadly three trends visible in this arena: first, nature writings that express the outdoor, intimate, positive observation, and relational reflection of the natural world; second, religious writings that illustrate the beauty, divinity, and sacrality of the natural world and also describe nature worship and mysticism; and third, eco-literature, which underlines the creative confluence of nature, culture, identity, people, and politics. Such trends have often been parallel to each other, even while having overlapping tendencies.

Historical research on nature writing in ancient India captures the nature narrations in the Rig Veda, the Ramayana, and the Mahabharata, and in the works of Sanskrit poet Kalidasa.[16] There have been several nature writers in Mughal,[17] British, and independent India.[18] In the recent past, writers like Salim Ali, M. Krishnan, Ruskin Bond, and Stephen Alter, organisations like the Bombay Natural History Society, and publications like the *Sanctuary* have enriched environment literature.[19] However, most of the themes of nature writing in India's intellectual and literary canon, inspired by traditions of romanticism and naturalism, have revolved around forest, wildlife, river, mountain, sea and natural landscapes, personal memoirs, conservationist love and passion, wilderness, rural life, and natural world as a place of pristine refuge. It has been remarked:

> Nature writing is an ambiguous phrase that is difficult to define.... Most of the time, however, it implies a scientifically informed yet literary approach to any subject in the natural world. Thankfully, that eliminates political and military memoirs, computer manuals, and books on spiritual wellness.[20]

Another important source of nature writing has been religious and spiritual commentaries.[21] In India, Hinduism has been a major source of such nature imagination. Rivers are considered sacred and are described through scenic, mythical, and ritual details. A physical and spiritual journey of India's most sacred river Ganga travels across snow-covered mountains, forested valleys, holy pilgrimages, myths, and traditions of temples and epics, revealing intimate connections between natural history and mystical experience. The great circumambulation of the sacred Narmada – a pilgrimage along the longest west-flowing river in the country – goes from source to sea and back, and in between takes a scenic route of historical and cultural sites of deities, temples, shrines, hills, forests, rivers, ghats, and villages. This is described as a complex religious exercise and a challenging nature trail. In another instance, the region Braj, the river Yamuna, and the Hindu deity Krishna become divine symbols of nature.[22]

The above two traditions have been so dominant that eco-literary writings, which combine the form and aesthetics of creative literature to portray physical and human environment interactions in multi-layered ways, have not been worked much in India. There have been a few environmental narratives in Indian creative literature from the perspective of humanities, expressed in the disciplines of history, geography, anthropology, and philosophy.[23] Over the past two decades, we have had some contributions in environmental literature.[24] These are literary collections on rivers or forests that remember our past contact with these elements, value nature preservation, and reflect on the complexities of the present.[25] Such writing has recently developed further, and captured some new environmental issues like receding glaciers and global climate change.[26]

However, Dalits and their vast range of literary products are missing in all the traditions of literature-environment studies. Poetries, stories, and memoirs composed by Dalits and anti-caste writers reveal meaningful attachments to nature and the outdoors, even though a separate strand of 'nature writing' has not been recognized in the weaves of Dalit literature. Yet, Dalit ecological experiences and expressions can be seen in different ways – as an intellectual exercise, a product of labour, a source of creativity, an environmental space that is interwoven with a politics of survival and struggle, a way of living with and understanding natural resources, a dialogue with nature where notions of existence and difference are played out to know the self and the other, and an arena where questions of 'casteized' and 'casted' beings can be extracted at the level of community and cultural production. I read three Dalit autobiographies through the lens of literature-environment studies and highlight how a framework of eco-criticism and eco-justice can enrich our understanding of Dalit eco-literary traditions.

Sharing Breath: Environmental Justice, Eco-criticism, Caste, and Dalits

Explaining our imaginative failure in the face of global warming, Amitav Ghosh refers to the novel *A River Called Titash*,[27] written in 1956 by Adwaita Mallabarman, who belonged to an impoverished caste of Dalit fisherfolk. The novel is set in rural Bengal, in a village on the shores of a fictional river, Titash. Adwaita (1914–51) grew up in his village community, later migrated to the city, and died of tuberculosis in an obscure hospital at a very early age. His novel (only work of fiction, published after his death) is characterized as autobiographical in a wider sense – a self-reflection as part of a community, its culture, and living environment. The novel begins with the following narration:

> Titash is the name of a river. Its banks brim with water, its surface is alive with ripples, its heart exuberant. It flows in the rhythm of a dream....
>
> But all the rivers are not similar in appearance and essence. How they behave with people differs and how people behave with them differs too.... The river there is a frenzied sculptor at work, destroying and creating restlessly in crazed joy, riding the high-flying swing of fearsome energy....
>
> Titash is just an ordinary river. No one will find its name in the history books, in any of the chronicles of national upheavals.... But its banks are imprinted with stories of a mother's affection, a brother's love, the caring of a wife, a sister and a daughter. That history perhaps some know, perhaps some don't. Still, that history is true and real.[28]

The novel's four parts have complex webs of narratives, throughout getting in and out of nature, with vivid descriptions of river; Malo fisherfolk;[29] a variety of fish, nets, and boats; vibrant community life and cultural spirit; formation of silt bed; beginning of wet-rice farming; and painful displacement of the Malos. We witness the shattered journey of a fisherman through nature, who attempts to claim his personal and social space. The Malo moves forward and backward in space and time for a new spatial identity. Along the river, a woman searches for self and her lost family. The life of a river is transformed beyond imagination. A restless boy expresses an endless hunger for knowledge of nature and education. A fisherman lives with anger against the exploitation by upper-caste traders and moneylenders and becomes intimate friends with Muslim peasants. The narrative's structure, dialogue, and lyrics are immersed in nature:

> The relationship between people, river, and seasons permeates events and images, metaphors and musings, songs and dialogues…. Daily life and festivals, events and reveries reflect the keenly perceived changes in nature. The river both evokes and affects lives and thoughts.[30]

The long quotes that I have cited as an epigraph for this section demonstrate how Dalit self-narrations can be extended to reflect environmental justice. Environmentalism has often focused on the role of anthropogenic in articulating contemporary environmental issues and agendas, which is at times based on the simple scenario of 'human versus nonhuman'. However, even without denying the serious human impact on environment, anthropocentrism – 'the destruction and degradation of nonhuman ecosystems by human beings'[31] – has been seriously critiqued for not recognizing gender, race, and ethnicity in shaping human–nature interactions. Ideas of environmental justice and the emergence of African American, Native American, and Dalit movements since the early 1980s offered a channel to articulate that environmental discrimination on the basis of colour, race, caste, origin, income, cultural difference, and unequal distribution of environmental risks and benefits must be integrated into environmental politics. Dalit human rights movements in India raised issues of caste discrimination, domination, and access in land, water, forest, space, and landscape. Intersections between human rights and environmental justice movements in India further emphasized social justice, equality, and diversity in environmental arenas. In this context, Dalit autobiographies can also be read as a literature of environmental justice. Julie Sze, who has extensively worked on environmental inequality and justice, in her reading of Japanese American author Karen Yamashita's novel *The Tropic of Orange*, states:

> Literature offers a new way of looking at environmental justice, through visual images and metaphors, not solely through the prism of statistics. This new way of looking references the 'real' problems of communities struggling against environmental racism, and is simultaneously liberated from providing a strictly documentary account of the contemporary world.[32]

Post-colonial critiques – 'critical discourses which thematize issues emerging from colonial relations and their aftermath, covering a long historical span (including the present)'[33] – have engaged with the ecological question, natural environment, and our relation to it through specific social histories. Such historical examinations, underlining links between the empire, European global expansion, and environment, also made it 'possible to perceive a new kind of concern for the environment emerging in the post-colonial era, one attuned to histories of unequal development and varieties of discrimination, including, of course, racism and sexism'.[34] Frantz Fanon's works provided an impetus for postcolonial ecology by drawing a humanistic relationship between life and land and the emergence of resistance in life, as it engages with land.[35] Edward Said re-imagined land, territory, space, geography, dispossession, displacement, identity, and memory to question the control over people and place, not only in traditional 'geographic' terms but also in a broader epistemological sense.[36] Ideas of environmental justice that arose from such backgrounds have taken numerous forms in theory and practice. However, they have been broadly defined in the context of ownership of natural resources and distribution of benefits and burdens in a political community.[37] Dalit descriptions of environmental justice place caste and hierarchy as central elements in assessing environmental issues in a natural and social order. Simultaneously, they expand the horizons of cultural products, including the languages, through which we can explore a new environmental universe.

The developing discipline of eco-criticism – study of relations between literature and the physical environment[38] – also provides critical tools to locate Dalit literature within nature. It helps in integrating Dalit cultural representations and sociopolitical struggles into a 'natural' space and place. Since the 1980s, literary scholars around the world have defined and applied eco-criticism widely as a literary and cultural space to decipher nature. By reflecting upon nature in literature, they track environmental ideas in partially concealed cultural spaces, viewing it as a response to environmental crisis. They call for a cultural change by broadening human conception of a global community to include non-human forms and the physical environment. A pioneer in this field, Scott Slovic characterizes this development in four waves that have beginnings but no end-dates. The first wave, beginning in the 1980s, had its prime concern in the protection of 'natural environment' – non-human nature and wilderness – from humans. The second wave, beginning in the mid-1990s, moved away from nature writing and included rural and urban landscapes, with multicultural voices and multiple genres. The third wave, beginning in 2000, had a diversity of voices from various national and ethnic cultures to understand different human relationships to the planet. And the fourth wave focuses on the fundamental materiality of environmental things, places, process, forces, and experiences of the human body and the natural world.[39] Dalit autobiographies can expand the arena of eco-criticism, as they narrate their broken ties with the earth, where histories of caste injustices and living within hierarchies and pollution make them the 'Other' in everyday environment. Their living experiences bring forth different meanings of space, touch, smell, air, fire, water, and other environmental resources. They thus provide a counter-narrative to the universal meanings of environment.

The 'Eco-criticism of the Global South'[40] is a new coinage that further challenges the hegemony of political, economic, and social powers, and locates the intersection of nature and culture in multi-locations, beliefs, and practices. For example, some Sri Lankan Anglophone fiction writers have critiqued elitist histories of plantations that have negated subaltern perspectives. Swarnalatha Rangarajan writes that *Heaven's Edge*, a 2002 novel by Romesh Gunesekera, offers a critique of plantation monoculture and the devastated ecologies it produces. Set in a post-apocalyptic island, the novel depicts the narrator's attempt to unearth the palimpsestic history of labourers and land in an abandoned tea plantation that he encounters. Likewise, Jean Arasanayagam's stories give voice to these suppressed narratives and span a variety of topics, ranging from the plight of indentured plantation workers in the nineteenth century to the horrors of paradise tourism and the plight of migrant workers in contemporary times.[41]

Similarly, the works of Chinese poet Zheng Xiaoqiong and Pakistani writer Uzma Aslam Khan underscore a complex ecological web of migration, labour, deforestation, globalisation, and economic development, which is markedly different from mainstream narrations. In the following section, three Dalit autobiographies are located in the domain of casteized nature and environment, and highlight how Dalit writings can bring to us a new universe that can enrich our ecological, literary, and cultural theories.

The Words of Dalits in the World of Nature

Bama's *Karukku* chronicles her wounded life from childhood to early adulthood as a nun, in doing so eschewing linear or chronological order. Her Tamil, Dalit, woman, and Christian identities criss-cross each other as she is subjected to severe caste oppression. She constantly struggles for her dignity and identity in a community. Manohar Mouli Biswas's *Surviving in My World* is a collage of memories of growing up as a Dalit boy in Bengal – poor, illiterate, powerless, sick, and isolated in village life. At the same time, his single-minded perseverance of getting an education against all odds pulls him out of his subjugation and earns him respect in a caste society. Siddalingaiah's *Ooru Keri* deals with his childhood and youth in rural and urban India. It chronicles the contempt, humiliation, and violence that he encountered, and his personal, political, and cultural emergence from it due to his education, which gave him great confidence. These autobiographies do not offer a comprehensive account of Dalit eco-literature. Yet they are representative enough and carry within them a diversity of texts and topics that signify Dalit ecological insights. In different ways, they show how caste and nature are intimately and inextricably interwoven in India and why nature is critical in shaping Dalit lives. Living with nature, these autobiographies constantly negotiate with, and challenge, caste domination, while simultaneously articulating an environmental imagination. They not only enrich eco-literature but also weave it together with environmental justice. I will focus on some nature markers in these works to demonstrate my central premise.

Colours of Nature

Alison Deming and Lauret Savoy underscore that the narrations of many communities are missing from mainstream nature writings. They raise the question, 'Why is there so little "nature writing" by people of color?' They locate its answer in the way we have been defining mainstream nature writing and writers. Going through the diverse writings of African American, Arab American, Asian American, Latino/Latina, and Native American people, they argue that

> the experiences of all people on this land are necessary stories, even if some voices have been silent, silenced, or simply not recognised as nature writing. What is defined by some as an edge of separation between nature and culture, people and place, is a zone of exchange where finding common ground is more than possible; it is necessary.[42]

Dalit life narratives, too, express an intimate and everyday connection with nature. Their narrations are also manifestations of ecological spaces, with all their beauty, diversity, colour, and complexion. There emerges a biosphere in its various forms of liveability, which has a daily affiliation with the individual and the community, and with the human–non-human population of the planet.

Bama vividly describes the natural beauty of her village. Her story contains mountain ranges and their peaks, rocks, woods, fields, foxes, rains, rainwater leaping down the mountain slopes and filling the streams that encircle the village, lakes and ponds standing side by side, and abundant fish in stretches of water that people can catch easily by placing earthen pots at strategic places. Recounting the splendour of natural beauty, she perceives nature as an active entity, transforming and responding to her being as a human. Nature gains consciousness and takes different strides, from simple organisms to intimate entity, from vertebrates to human society. This creates a nature–self–society web, with all parties active, although the forms in which she manifests such activity are different from dominant nature writings. She writes:

> When we used to go out in the early morning to relieve ourselves, a bright red sun, huge and round, would wake up in the east and climb into the sky. It would make its way, peering between the trees, glowing, its light spilling and sparkling. And in the same way, at evening time, when it went and dropped through the mountains, all the fields around would be luminous with a yellow light.[43]

Such memories of nature appear in her biography through hundreds of references, which are reminiscent of her everyday village life. Rich, varied, and detailed narratives of rivers, ponds, wetlands, forests, mountains, and land reflect an artist's imagery and creativity. They are real and colourful places that are to be enjoyed, celebrated, and communicated. This gives rise to a new possibility of existence and development of a free biosphere, which is a product and result of its own evolution and has to be lived with passion. Intimately connecting with nature, Manohar Moul Biswas experiences an 'otherness' in his natural worlds. His spatial, social, and economic situation also makes him realize his own exclusion from the village, garden, and home. The tension between inclusion and exclusion within the realm of nature and his self is all encompassing:

> The wetland was deeply embedded in my mind. It was so massive that by standing in a village along one of its edges, the villagers on the other side appeared minute. The trees appeared short; the golden hued ripe rice fields were like courtyards neatly smeared with cow dung in the early morning. In someone's huge sesame garden flowers had blossomed on each bush in clusters.... The small mango or betel nut gardens in some people's houses were standing with their heads raised in a dark cloudlike canopy.[44]

Living amidst these forests and lands, Biswas continues to express his physical, personal, and emotional attachment with them. He has seen them through changing seasons and times, and experienced multiple threads of his life and identity within and around them. He goes on to say:

> When winter spread its wings to announce its arrival, then the appearance of the nal forest [*Phragmites Karka*] changed. The tips of the nals would flower, appearing as if wearing numerous white chamors – like feathers of a fly whisk – on their heads and they were dancing in the wind.[45]

Descriptions of nature abound in narrations of rain, river, fish, forest, hill, land, and agriculture. They celebrate the intrinsic beauty of nature and humankind's interaction with her, reflecting a unity between individual, society, and nature. The poetic language and literary expression of Dalit autobiographies sometimes adopt nature imaginary through familiar idioms and forms, which have many things in common with 'mainstream' nature writing and literature in India. However, there are other distinct elements as well in their life narratives, on which I reflect now.

Caste Nature

Nature is seen as natural, common, and inherent. However, nature is not devoid of caste. Dalit autobiographies constantly underline their everyday ecological burdens in a marked hierarchal order. Images of land animate caste anxieties around labour, blood, and bondage. In dry regions, Dalits must often sacrifice their lives to recharge ponds and water resources. From village to city, and temple to school, caste metaphors of pollution, impurity, and dirt dominate places and spaces, through imaginaries of dangers posed by the presence of Dalits. Forests can be a haven or a hell for Dalits. Nature, entwined with fear and violence, horror and hardship, bloodbath and war, makes environmental experiences of Dalits distinctive and different. A river is not just a sacred, pure, religious, and spiritual body. It is not only a place for devotees and their prayers but also reminds Dalits of their pain, body blows, challenges of boatmen, waves causing havoc to boats, and struggles for survival. Biswas writes in his autobiography:

> Bhairab is a frail river. We called it ... a female river. Just as the trading boats of the paddy sellers floated down Bhairab, so did the trading boats of the jute sellers.... Someone would light a stove under the chhoi and cook. In trying to light the fire with straw his face would be lashed by smoke when the wind blew

from the opposite direction.... Waves broke on the banks. The waters of this river were muddy, invoking compassion, like looking at the pale face of a woman in pain. It was as if it were carrying a pain somewhere in its body.[46]

Bama remembers the mountain jungles where she collected firewood, and the hard and cruel life of a low-caste woman. She recounts how she was forced to survive on forests, and how she had to give some money to the 'guarder' or forester to allow her to collect firewood. Twigs and thorns scratched and tore her face. Sometimes her skin started bleeding, or her hair got entangled in the branches, almost splitting her skull apart. She often pushed, shoved, and crawled her way through bushes and briars. There is a specificity of casteized nature, which is different from the universal question of accessing nature. Land, forest, river, water, agriculture, village, city, occupation, and food have been important sites for impositions of hierarchies of caste. However, the interactions of Dalits with casteized nature have several layers that do not often fit into any neat categories of natural and human or pristine and prison. Several images of physical and natural landscapes illustrate how places of refuge have not only produced a deliverance from bondage to freedom but, more importantly, a transformation from rootlessness to rootedness for the community. They show how Dalits longed to change their landscape, where they had to live in poverty and drudgery, into a liveable place. In his autobiography, Siddalingaiah remembers his city-space:

A big drain lay in front of our house. It was the holy Ganga to us. It would fill up and flow during the rainy season. Watching the flowing drain was great fun. A variety of objects came floating by. Coloured paper, clothes, old cradles, baskets, palm fronds, fish and snakes float by. Once a corpse came floating.... Even when it stank, our affection for it did not wane.[47]

The drain becomes a metaphor for a space that supports everyday life-making practices, a space that is for Dalit public use, a space that spurs the movement of their bodies and thought. Space has an overarching influence on Dalits' lives and thinking, and in their quest for equitable distribution of physical and social arenas. The location and space of a drain symbolizes a social ghetto, historically produced and reproduced. However, Dalits frequently move in and out of this space. Siddalingaiah has numerous narratives about the drain in his locality:

People from the factory on the other side dumped mounds of black earth into it. Small pieces of bronze and copper could be found in these mounds. We would descend into the drain and gather them by digging up the mounds.... For us the drain had become an inexhaustible trove that could earn us candy.... We hoped the drain could throw up even more precious things. So we would descend into it and walk for a long time. As it was a sewage line, it would be like a tunnel journey.[48]

Nature, casted in caste, mirrors the subjugated status of Dalits. It is imagined and re-imagined, worked and reworked in Dalit environmental and life narratives. They see themselves in water, river, land, and agriculture, which are marked by hierarchy, pain, and powerlessness. Manohar Mouli Biswas defines himself through nature, while also pointing to its social order:

The people born in nature, lived in their own way and even died in their own way. The name of this history of life and death is prisnika – growing up like the water hyacinth and dying like it, uncared for. I was born into such a community and that is how I grew up in my deprived childhood.[49]

Dalit autobiographies starkly question the dominant romanticism of nature, as they regard violence and discrimination as parts of a malignant 'natural order'. Nature is an arena where questions of 'casteized' and 'casted' beings can be extracted at the level of self, other, community, and cultural production. For Dalit writers, considering nature as some kind of a superior entity cannot be more wrong. After all, nature also brings decay, sickness, and death. Nothing can be more disturbing than to make nature an object of reverence and worship, even before one starts addressing the monstrous social injustices that are accepted as a consequence of blind faith in nature. Nature is a source of wonder and fear, for it will claim everybody's life in the end. Nature is also a major distributor of existential injustice. No one asks to be clever or beautiful, ugly or disabled, mentally unstable or emotionally unbalanced. These injustices are bad enough without human societies adding to them and imagining that some human beings are by their birth, fate, and destiny beyond the purview of human society.

Labouring with Nature

The absence of labour within the frame of mainstream environmentalism has been critiqued.[50] It has been asserted that a focus on work can improve environmental discourses about the natural world and give a better sense of the daily lived ecological experiences of the labouring population.[51] An eco-critical dimension of labour takes us to different environmental questions:

'What does it mean when work, rather than leisure, is your central ecological experience? What does it mean when work is compounded by the inconvenient history of enslavement?'[52] Dalit autobiographies place their labour at the centre of explaining relations between society and nature, and work through the vantage points of caste, labour, and nature. Their labouring bodies work over natural resources. Their labour is bound by traditional, old, and hierarchical conditions of human–nature interactions. Their degraded labour is a source of unequal and unjust distribution and consumption of fruits of nature. Autobiographies of Dalit women and men are often 'earthly' in nature. They breathe the air of the earth, eat earthly food, drink earthly water, and live with earthly labour. Biswas describes that we are 'people of mud and water', 'natural warriors of physical labour', 'hardworking people by birth', and for us labour is 'another name for life'.[53] As labouring people, despite their capacity to adapt to ecologically and socially hostile environments, Dalits find themselves in difficult situations. Biswas continues: 'Exposed to the opposing rage of nature, in some years people became helpless against the hardship they faced. They took any manual work in order to tackle poverty.'[54]

Nature is socially embedded in, and mediated through, caste, which in turn shapes and influences Dalit conditions of labour. The process of being a Dalit is also a process of branding stigmatized labour as part of nature, a 'natural' labouring technique that belongs to the body of the Dalit. Dalits' hands become their tools, through which they labour in nature. Dalit autobiographies are testimonies of their hand tools, which are used for agriculture, fishing, boating, hunting, and other manual occupations in rural areas. When they migrate to cities, their hands remain their instruments of labour, leaving deep scars on their caste-marked bodies. There are no threads in Dalit life narratives of an ecological idealization of labour–nature interrelationships. Instead, they offer a critique of a simple, natural, and harmonious relationship between labour and nature. In *Ooru Keri*, there is a long account of land and labour:

> One day, as we stood on the squat walls calling out to our parents, we noticed something strange. A man had fastened a yoke onto the shoulders of two others, and was ploughing Ainoru's fields. It was amusing to watch the two men trundle on like bullocks, while the third followed them swinging a whip and making them plough. A strange agony gripped me the moment I realized that one of the men carrying the yoke was my father.[55]

Even after migrating to the city, Siddalingaiah's father continued to do back-breaking jobs. At times, he was employed as a temporary worker in a factory;

at others, he was axing small logs into splinters at firewood depots, bundling and delivering them at homes.

Biswas remembers Dalits' labour in agriculture with a sense of pride. The world of agriculture writes its history not just through land and environment but also through caste and Dalit labour, which even time is powerless to erase. He writes:

> My father, grandfather, and all my illiterate forefathers were great agriculturalists of Bengal. About 80 percent of our community produced the harvest in the wet earth and fed the whole of Bengal. Why should I look down on this occupation? Why consider the belittlement of the Aranyakas mentioned in the Sam Veda songs? The middle-class, upper-caste babus were not attached to this occupation.[56]

Bama describes the hard and incessant labour of her mother, grandmother, people of her community, and herself from sunrise to sunset. They had to sweep the streets, dredge and clean the drains, search for thorns used for fences, pick up palmyra and coconut palm stems and fronds for fuels, collect fresh cow-dung and pat it into flat cakes for burning, pull up the groundnut crop, clean and sort the pods, collect firewood from the mountain forests and take them to the market, and work as bonded labour. Labour thus figures as a central element of Dalit lives through the never-ceasing interactions with the surrounding environment. This everyday physical work and its acute awareness have significant bearings on Dalit understandings of the natural world, as they are often in concordance with the form and content of their labour activity.

It is precisely through the process of labour that nature is also 'humanized' in Dalit narratives. This process mostly takes place outdoors, through the agency of an individual or a collective. Nature remains a hostile environment for Dalits; yet they also see the distantness and proximity of the stars, hear the whispering of the trees, and appreciate the strength of the fertile land. Bama says that women and bare-bottomed toddlers then sing: 'As they planted out paddy seedlings, or weeded the fields, or harvested the grain, they worked to the rhythm of their songs.'[57]

In Dalit life narratives, labour also gets socialized and converted into their collective bodies to transform nature and their 'natural' situation. Through the impact of organized labour, the alterations in nature take a positive turn. It is liberating in so far as it initiates a call for dignity and designs and creates a language of protest. Biswas narrates that people living on the banks of the river Kali, mainly Dalits, needed to build a bridge across the river for public goods. They had to plan, design, dig mud, and arrange boats, bamboos, and straws on a massive scale. The outcome, he says, was historic:

> Such a situation where the people of Chitra and Kali river banks united for collective physical labour was a hoary tradition. It was difficult to find anything comparable to their capacity of settling big tasks with physical labour and their entrepreneurship.[58]

In short, narratives of labour and nature abound in Dalit autobiographies. Their conceptions of life, earth, and nature centre on the value of labour, which remains an overarching concern in their eco-literary telling. Labour is intrinsically tied to livelihood, the environment, and character formation, as it creates fresh landscapes and cultivates new virtues.

Animal Nature

Animal tales of marginal communities have been interpreted differently in eco-criticism, in terms of their portrayals of community as part of a unified universe of all creatures or in depictions of social relations as inscribed in nature.[59] Animals, animal lives and behaviours, humans and animals, and occupations and animals are crucial reference points in Dalit autobiographies. Animals circulate across their life narratives and enter into socio-spatial and purity–pollution relations. Animals and their 'animality' construct a self-identity – a degraded low-caste untouchable – with natural history and geography. Dalit lives are often compared with animals – they are often treated like them. Animals thus become markers of social relations, power formations, and difference. Dalit interactions with animals reveal a paradox. Dalits as humans live and work closely with animals in the physical space, but in their inner cultural space, they often view animals as 'others', to be discarded and displaced. There is also a specific caste–Dalit model, as they are forced to deal with dead animals in a pre-determined natural–social world.

Like many other Dalits, Bama and Siddalingaiah grew up with domesticated animals. They grazed cattle in the village commons and the mountain forests. They could 'unite' the animals during cattle grazing, milk the cow, make an effigy of a calf to yield milk from the mother cow, rear the pigs, hunt the dogs, and trap the wild pigs, foxes, and rabbits. This was a part of their everyday life. One day, Siddalingaiah went along with his cousins to graze his cows and sheep. He was very fond of a particular sheep who was pregnant. He narrates:

To my amazement, it gave birth in my hands. The little new-born started hopping around. I called out to the others. We made the little sheep drink milk from its mother's udders. To ensure that it did not harm the baby, I placed my right palm in its mouth. I don't know how it felt, but it sank its teeth into my hand. Blood streamed out.... To this day the scar shows on my fingers.[60]

However, animals and the idea of animality acquire opposite connotations when Dalits define themselves vis-à-vis their caste-ridden living situations. To be an animal has been understood as not-human, to be a commodified entity, subject to the whims of humans. Hardly any Dalit autobiography goes without a statement proclaiming that they are forced to live like animals. Biswas characterizes his community 'like the common chunoputi fish, stayed back in our motherland, primarily because of sheer helplessness'.[61] Bama compares the whipping of her people 'like they whip animals until they can neither see or breathe'.[62] A caste society is characterized by Dalits as akin to an animal society, with Dalits aspiring for a human society. Such views indicate Dalit perceptions of 'natural' social relations. There is no simple, straight-jacketed thinking about harmony in nature between humans and animals, as they are inscribed in both nature and social relations.

Dalit autobiographies meticulously depict specific animals, immersing readers in the intricacies of their food habits. Through this exploration, these autobiographies skillfully intertwine narratives that dip into life experiences, culinary practices, and the interconnected world of surrounding animals. Such explorations provide layered readings of how animal foods are collected, prepared, and consumed by Dalits; how animals like cows and pigs have been associated with Dalits and their food in specific ways that are determined by their caste; and how representations of animal-related foods are closely intertwined with power, class, caste, and gender. Meat remains an important food source for many Dalits, and particular dishes hold social connotations, as they are at times associated with specific delicacies and special events. Some dishes and meats are tied to Dalits' oppressive pasts, though they have been adapted and have evolved into new tastes and cuisines. Biswas intimately recounts how pigs – which he describes variously as grey-black, some blackish in colour, reddish ting on their breasts, a more reddish long mouth, thick and longish nose-flutes and small eyes – were a rare and cherished food. Food preparation was a space of imagination, ingredients, and creativity:

Buying and eating pork was a matter associated with merriment.... The buyers, at least four to five of them, tying a rope round its legs and hanging it on their shoulders. And what did Subal and Nakhye do on such an occasion? They would keep their house, its courtyard, the cowshed and every nook and corner neatly smeared with cow-dung-water. It would get the form of a temple in cleanliness.... Then they washed the dead animal with hot water. Its body would gradually shine.[63]

Bama describes the delicacy – 'a good meal' – of cow meat, even with its limited access and constraints of cooking: 'We would be given a little of the meat ... the meat was just boiled with a touch of salt.'[64]

Dead animals are a crucial marker in these autobiographies, which underline human–animal distinctions. Dead animals mean different things to different people in a caste society. For the dominant-forward castes, they symbolize deadly pollution; for 'untouchables', they are part and parcel of their birth-bound occupation, where it is their 'duty' to remove the carcass. It signifies a forced, polluted, and traditional occupation; a source of food, meat, and income; and a symbol of death, decay, blood, and smell.

Animals are conceptualized differently in Dalits autobiographies. The animate environment, of which humans and non-humans are part, can be torn apart in human versus animal scenario, under specific conditions. Knowing the animals intimately, conceiving 'our' humanity by contrasting it to the animality of the 'other', dreading dead animals, yet recognizing their relationship to food and occupation, makes animals a fascinating focus to illuminate Dalits' different ecological world reflected in their literature.

We can try to capture and extrapolate Dalit environmental consciousness and narrations through their autobiographies and literary texts. They underline how nature and caste are inextricably intertwined in Dalit narratives, where environment is not a pristine, innocent player but plays a critical role in the stigmatization of Dalits and in their social exclusion, while also being a significant source of their everyday livelihood. In this context, four dimensions have been focused: the depth and breadth of Dalits' intimate encounters with nature and their celebration of it through rich and colourful expressions; the overarching presence and the painful experience of caste in nature; the central role of labour in understanding Dalit relationship to nature in their everyday lives; and the conjunctions between the consumption of certain animals and Dalit identity formation. While there are many other nature markers that animate from Dalit eco-literary writings, this selective survey reveals an unexplored aspect of Dalit literature and autobiography. Dalit autobiographies offer complex narratives, and in their eco-spaces we can find multiple foci. These nature markers are embodied in their life narratives, appearing at several intersections of their personal, political, social, and cultural journeys. While Dalit autobiographies have been widely recognized and perceptively analysed by scholars, there is a strong case to argue that these texts must also be read in the context of their ecological universe. When the world of environmental justice and eco-literature is viewed through Dalit writings, it acquires richer and varied meanings.

The troubling question of why there has been no mention of Dalit literature in the wider canon of nature writing interrogates its very definition in the past and the present. Dalit stories, images, metaphors, and anecdotes challenge the lenses of beauty, recreation, and spirituality through which nature has been dominantly expressed. Dalits' place on the earth is constantly on the move, and through their storytelling of the present space, they also attempt to transcend it. It is no coincidence that the self-narrative of Siddalingaiah ends with songs of Holeyas and Madigas,[65] whereby he locates himself in a new spatial relationship with the earth and its natural objects. In the final strokes of Manohar Mouli Biswas, everything that he loves – the Bhairab river, the banks of the Kali river, the flowers of the field, the common fish – merge into one, and he continues to flow like a quiet river with no use of force and no holding back. The lively Bama sums up her autobiography through three prominent figures of the environment: tree, fish, and bird, which signify her search, transformation, and achievement of the self.

In the next essay, I will analyse other major trends of Dalit eco-literary tradition, that is, mythological, oral, and anecdotal, which are crucial to understand the ways in which Dalits have engaged ecology. Their stories and characters are not only limited to domination and oppression. Dalit folklore such as Dina-Bhadri takes on dynamic and shifting storylines that weave together narrations of labour, ecological resources, and struggles against Brahmanical tyranny and landlordism. I especially stress how the folktale can morph and shift its ending according to context, and depending on the particular situation and contemporary challenges of the Musahars who are narrating it.

Notes

1. Bama, *Karukku* (New Delhi: Oxford University Press, 2012).
2. Manohar Mouli Biswas, *Surviving in My World: Growing Up Dalit in Bengal* (Kolkata: Samya, 2015).
3. Siddalingaiah, *Ooru Keri* (New Delhi: Sahitya Akademi, 2003).

4. A Dalit caste in West Bengal state, earlier known as Chandala or Chandal, a term usually considered a slur.

5. I have emphasized elsewhere how caste and nature are intimately connected, while comparing Dalit meanings of environment to ideas and practices of neo-Brahmanism and certain mainstream environmental thought: Mukul Sharma, *Caste and Nature: Dalits and Indian Environmental Politics* (New Delhi: Oxford University Press, 2017).

6. There are a number of volumes documenting such Dalit writings. To name a few, K. Satyanarayana and Susie Tharu, eds., *No Alphabet in Sight: New Dalit Writing from South India, Dossier 1: Tamil and Malayalam* (New Delhi: Penguin Books, 2011); K. Satyanarayana and Susie Tharu, eds. and intro., *Steel Nibs Are Sprouting: New Dalit Writing from South India, Dossier II: Kannada and Telugu* (Noida: Harper Collins, 2013); Sharmila Rege, *Writing Caste/Writing Gender: Narrating Dalit Women's Testimonies* (New Delhi: Zubaan, 2006).

7. A large body of research publications exist on this. To name a few, Rosalind O'Hanlon, *Caste, Conflict, and Ideology: Mahatma Jotirao Phule and Low-caste Protest in Nineteenth-Century Western India* (Ranikhet: Permanent Black, 2002); Namdeo Dhasal, *A Current of Blood* (New Delhi: Navayana, 2007); Ramnarayan S. Rawat, *Reconsidering Untouchability: Chamars and Dalit History in North India* (Ranikhet: Permanent Black, 2012).

8. For example, D. R. Nagraj, *The Flaming Feet and Other Essays: The Dalit Movement in India* (Ranikhet: Permanent Black, 2010); Manu Bhagavan and Anne Feldhaus, *Claiming Power from Below: Dalits and the Subaltern Question in India* (New Delhi: Oxford University Press, 2008).

9. For details, see Pradeep K. Sharma, *Dalit Politics and Literature* (Delhi: Shipra, 2006).

10. These debates are well captured by Sharankumar Limbale, one of Maharashtra's pre-eminent Dalit writer-activists. Sharankumar Limbale, *Towards an Aesthetic of Dalit Literature: History, Controversies and Considerations*, trans. Alok Mukherjee (Hyderabad: Orient Longman, 2004).

11. S. Anand, ed., *Touchable Tales: Publishing and Reading Dalit Literature* (Chennai: Navayana, 2003).

12. Raj Kumar, *Dalit Personal Narratives: Reading Caste, Nation and Identity* (Hyderabad: Orient Blackswan, 2010), 150.

13. Debjani Ganguly, 'Pain, Personhood and the Collective: Dalit Life Narratives', *Asian Studies Review* 33, no. 4 (December 2009): 429–42.

14. M. S. S. Pandian, 'Writing Ordinary Lives', *Economic and Political Weekly* 43, no. 38 (20 September 2008): 35–40.

15. Rege, *Writing Caste/Writing Gender*, 15.
16. Rajani Jairam, 'Ecological Concerns in Mahabharata', *IOSR Journal of Humanities and Social Sciences* 21, no. 5 (May 2016): 63–65.
17. Shireen Moosvi, 'Environmental Concerns in Mughal Era', *Journal of History and Social Sciences* 1, no. 1 (July–December 2010): 1–4.
18. Shashank Kela, 'Where the Wild Things Are Not: The Curious Absence of Contemporary Nature Writing in India', *The Caravan* (April 2018), http://www.caravanmagazine.in/reviews-essays/absence-contemporary-nature-writing-india, accessed on 22 June 2018.
19. Tara Gandhi, ed., *A Bird's Eye View: Collected Essays and Shorter Writings of Salim Ali*, vols. I and II (Delhi: Permanent Black, 2006 and 2007); Ramachandra Guha, ed., *Nature's Spokesman: M. Krishnan and Indian Wildlife* (New Delhi: Oxford University Press, 1997); Ruskin Bond, *The Book of Nature* (New Delhi: Penguin India, 2008).
20. Nature writer Stephen Alter talks about the past and the present of nature writings, along with a collection of nature writings. Stephen Alter, ed., *Writing Outdoors: A Natural Reader* (New Delhi: WWF-India, 2014), 5.
21. Diana L. Eck, *India: A Sacred Geography* (New York: Three Rivers Press, 2012).
22. David L. Haberman, 'River of Love in an Age of Pollution', in *Hinduism and Ecology: The Intersection of Earth, Sky, and Water*, ed. Christopher Key Chapple and Mary Evelyn Tucker (New Delhi: Oxford University Press, 2000), 340.
23. Upamanyu Pablo Mukherjee, *Post-colonial Environments: Nature, Culture and the Contemporary Indian Novel in English* (New York: Palgrave, 2010).
24. For example, Prerna Singh Bindra, *The Vanishing: India's Wildlife Crisis* (New Delhi: Penguin Books, 2017); Cheryl Colopy, *Dirty, Sacred Rivers: Confronting South Asia's Water Crisis* (New Delhi: Oxford University Press, 2012).
25. Amita Baviskar, ed., *Waterlines: The Penguin Book of River Writings* (New Delhi: Penguin Books, 2003), xii.
26. The acclaimed novelist Amitav Ghosh deals with global warming and climate change through literature. Amitav Ghosh, *The Great Derangement: Climate Change and the Unthinkable* (New Delhi: Penguin Books, 2016).
27. *Titash Ekti Nadir Naam* is a Bengali novel. The English translation of the novel was published in the 1990s. Adwaita Mallabarman, *A River Called Titash*, trans. K. Bardhan (Berkeley: University of California Press, 1993). The novel was adapted into a film in 1973 by the Bengali filmmaker Ritwik Ghatak. I will focus on *Titash* in a separate chapter.
28. Mallabarman, *A River Called Titash*, 10, 26.

29. A Dalit caste in West Bengal.

30. Adwaita Mallabarman, 'Introduction', in *A River Called Titash*, 4.

31. Sylvia Mayer, 'Introduction', in *Restoring the Connection to the Natural World: Essays on the African American Environmental Imagination*, ed. Sylvia Mayer (Hamburg: LIT VERLAG Munster, 2003), 3.

32. Julie Sze, 'From Environmental Justice Literature to the Literature of Environmental Justice', in *The Environmental Justice Reader: Politics, Poetics and Pedagogy*, ed. Joni Adamson, Mei Mei Evans, and Rachel Stein (Arizona: University of Arizona Press, 2006), 163.

33. Ella Shohat, 'Notes on the Post-Colonial', *Social Text* 31/32 (1992): 99.

34. Anthony Vital, 'Toward an African Ecocriticism: Postcolonialism, Ecology and Life and Times of Michael K', *Research in African Literatures* 39 (Spring 2008): 90.

35. Stephanie Clare, 'Geopower: The Politics of Life and Land in Frantz Fanon's Writing', *Diacritics* 41, no. 4 (2013): 60–80.

36. Joel Wainwright, 'The Geographies of Political Ecology: After Edward Said', *Environment and Planning* 37, no. 6 (2005): 1033–43.

37. Andrew Dobson, *Justice and the Environment: Conceptions of Environmental Sustainability and Theories of Distributive Justice* (New York: Oxford University Press, 1998).

38. Cheryll Glotfelty, 'Introduction', in *The Ecocriticism Reader: Landmarks in Literary Ecology*, ed. Cheryll Glotfelty and Harold Fromm (Athens: University of Georgia Press, 1996), xviii.

39. Scott Slovic, 'Seasick among the Waves of Ecocriticism: An Inquiry into Alternative Historiographic Metaphors', in *Environmental Humanities: Voices from the Anthropocene*, ed. Serpil Oppermann and Serenella Iovino (London: Rowman and Littlefield International, 2016), 104–06.

40. This is also the title of a book, representing eco-critical voices of the subalterns: Scott Slovic, Swarnalatha Rangarajan, and Vidya Sarveswaran, eds., *Ecocriticism of the Global South* (Lanham: Lexington Books, 2015).

41. Swarnalatha Rangarajan, *Ecocriticism: Big Ideas and Practical Strategies* (Hyderabad: Orient Blackswan, 2018), 103–04.

42. Alison H. Deming and Lauret E. Savoy, *Colors of Nature: Culture, Identity, and the Natural World* (Minnesota: Milkweed, 2011), 6–7.

43. Bama, *Karukku*, 4.

44. Biswas, *Surviving in My World*, 18.

45. Ibid., 62–63.

46. Ibid., 53.

47. Siddalingaiah, *Ooru Keri*, 29.

48. Ibid., 29.

49. Biswas, *Surviving in My World*, 48.
50. Richard White, "'Are You an Environmentalist or Do You Work for a Living?": Work and Nature', in *Uncommon Ground: Towards Reinventing Nature*, ed. William Cronon (New York: W. W. Norton & Company, 1995).
51. Mart A. Stewart, 'Slavery and the Origins of African American Environmentalism', in *'To Love the Wind and the Rain': African Americans and Environmental History*, ed. Dianne D. Glave and Mark Stoll (Pittsburgh: University of Pittsburgh Press, 2006).
52. Kimberly N. Ruffin, *Black on Earth: African American Ecoliterary Traditions* (Athens: The University of Georgia Press, 2010), 27–28.
53. Biswas, *Surviving in My World*, 56.
54. Ibid.
55. Siddalingaiah, *Ooru Keri*, 2.
56. Biswas, *Surviving in My World*, xx.
57. Ibid., 63.
58. Ibid., 22.
59. Kimberly K. Smith, *African American Environmental Thought: Foundations* (Kansas: University Press of Kansas, 2007).
60. Siddalingaiah, *Ooru Keri*, 10.
61. Biswas, *Surviving in My World*, 84.
62. Bama, *Karukku*, 36.
63. Biswas, *Surviving in My World*, 50.
64. Bama, *Karukku*, 72.
65. Holeyas and Madigas are Dalit castes who are mainly agricultural labourers and artisans in the Karnataka, Kerala, Andhra Pradesh, and Tamil Nadu states of India.

3

Culture

'God of Humans'

*Dina–Bhadri, Dalit Folk Tales, and Environmental Movements**

Dalits' histories, myths, and memories offer invaluable insights into their ecological past and present. These narratives, which have been widely disseminated through oral tradition, print media, and various forms of communication hold immense popularity within marginalized communities. Embedded within these stories, songs, myths, and performances are depictions of nature, landscapes, struggles for natural resources, acts of sacrifice, labour, and pain. Often characterized by their collective and anonymous nature, they have assumed significant relevance in the Dalit community's ongoing struggles for land, water, and other resource rights. A counterpoint to hegemony,[1] such cultural forms represent dissident voices, reflecting Mikhail Bakhtin's notion of dialogics and heteroglossia,[2] and Stuart Hall's concept of 'oppositional' decoding, challenging 'negotiated' reading positions.[3] It may be equated with Raymond Williams's paradigm of 'dominant', 'residual', and 'emergent' cultural practices in constant interaction.[4]

In the Indo-Gangetic Plains of India and Nepal, folk tales, stories, and songs woven around the two Musahar brothers, Dina and Bhadri, thrive among the Dalits, particularly those belonging to that community. Dalits living in the bordering sub-regions of Bihar and Uttar Pradesh and Nepali Terai have used myriad voices and a diverse set of folk traditions and cultures – oral tales, stories, songs, music, ballads, performances, proverbs, riddles, theatre, dance, festivals, crafts, and idols – to celebrate their folk heroes. The Dina–Bhadri folklore is a frequently sung ballad in north Bihar, with performances of as many as fifty-two wars of the heroes, waged to protect the poor labourers from

* Originally published as Mukul Sharma, '"God of Humans": Dina–Bhadri, Dalit Folktales and Environmental Movements', *South Asian History and Culture* 12, no. 1 (2021). Reprinted by permission of Taylor & Francis Ltd, http://www.tandfonline.com.

exploitation by the rich landlords. The folk tale has helped in transforming the rich cultural capital of Musahars into 'political and developmental capital for the betterment of the community as a whole'.[5] The legend of Dina–Bhadri is also effective for the mobilization of the marginalized Musahar community 'because of its strong anti-feudal, anti-bondage, and pro-peasant characteristics'.[6]

A study of the influential figures of Dina–Bhadri can signify a confluence of streams and issues – a popular folk tale and its dynamic reproduction in a variety of forms; changing histories of past and present; entanglements of texts, traditions, and performances; everyday lived social experiences of Musahars while producing and reproducing these stories; and a Dalit politics of folklore. While the folk tale has been a subject of some scholarly works, in contemporary times, it has also acquired dynamic ecological meanings, which have been relatively understudied. Bringing together a Dalit folklore and its environmental symbolism, I describe here how everyday internalized memories, cultural beliefs, narratives, and motifs of Dina–Bhadri as ecological ancestors are increasingly inspiring Musahars to mobilize and assert their environmental rights and agency. Based on extensive fieldwork since 1995 in the villages of five districts of Bihar – Madhubani, Darbhanga, Saharsa, Muzaffarpur, and Samastipur – I focus on this compelling relationship between folklore, Dalit narratives, and environmental movements. Also known as Bhuiyas, etymologically meaning 'of the earth', Musahars' environmental movements in the north Indian river plains creatively and repeatedly draw from representations and engagements with the folk tale of Dina–Bhadri.

Officially designated as a Scheduled Caste in India, Musahars are one of the lowest castes, and are considered 'untouchables' in the Hindu caste system. The Nepali Terai has a number of 'untouchable' groups, including Musahars, who are among the poorest in the country.[7] According to the 2011 census, their official numbers are 1,391,000 and 126,018 in Bihar and Uttar Pradesh respectively,[8] and they are 234,490 in Nepal.[9] They have been mainly landless agricultural and migrant labourers, who are particularly visible in the Indo-Gangetic Plains, one of the most densely populated agricultural regions of South Asia. The ecological character of the region is determined by linkages between highlands of the Himalayan mountains and fertile lowlands of the plains, and temperamental rivers like Kosi, Gandak, Ganga, and Ghaghara (Karnali) that emerge from the deep valleys of the Himalayas.[10] Historically, a land of plenty in terms of rivers, water, agriculture, and food, it has been increasingly witnessing ecological problems like worsening floods and droughts, shifting river courses, shrinking rivers, water loggings, high dams and embankments, land degradation, and fragmentation.[11] At the same time, increasing discontent has simmered among the Dalits because of social discrimination and unequal access to rights and resources. Growing ecological impoverishment has come face to face with Dalits challenging power relationships, leading to eco-political battles over land and water.[12] Simultaneously, a vibrant and creative politics has helped in building a strong cultural identity among the Dalits, including Musahars. Rejecting hegemony of Brahmin-dominated cultural universe, they have enthusiastically found ways to connect with their 'untouchable' pasts, and reaffirmed their cultural memories through folklore, music, gods, goddesses, festivals, and celebrations to carve their environmental movements and lay claims over political–public spaces. The folk tale of Dina–Bhadri has acquired varied meanings, appearing all along their rural landscape, what I refer to as 'Dalitscapes' – in villages, land and agriculture, rivers and water, forests and mountains – protecting the 'self', Musahar community, and Dalits.[13] Through an amalgam of places and incidents, Dina–Bhadri are projected as labourers, warriors, protectors, saints, ancestors, and gods. At times they labour peacefully, at others they fight courageously against the landlord; they die combating wild animals, yet are instantly reborn; they may be down but are never out in struggles against high-caste oppressors. The essay explores such entanglements by focusing on the themes of Dalit folk tales, their everyday memories, and folk heroes as ecological ancestors.

Folklore, Dalits, and Dina–Bhadri

'Folk' can refer to any group of people who share common, linking factors, be it occupation, language and/or religion, and has some traditions which it calls its own.[14] Depending on the context and audience, folklore is an important system, language, or register that people use.[15] A brief historical review of South Asian folklore studies shows its diverse components. Epics and folk tales like Ramayana, Mahabharata, Upanishads, Hitopadesa, Kathasaritsagara, and Jatakas have drawn attention of anthropologists and folklorists. Since the early nineteenth century, Christian missionaries, British civil servants, and Western philologists began documenting informative accounts of folklore. Towards the end of that century, Indian scholars brought out anthologies on the subject, arguing that 'a nation reborn must be inspired by its folk-songs', and post-independence India has seen a flourishing of works on folklore in multiple disciplines.[16] While finding considerable influence of ancient classical epics on oral epic traditions of India, scholars have touched on how the latter are also reshaped by local histories, oral narratives, and Muslim, Dalit, low-caste, and gender identities of some of their performers.[17] There is a recognition that the newfound vigour of folkloric production in South Asia is 'inevitably political' in terms of foregrounding issues of the self, and its narrative strategies link texts to social life and past to present.[18]

Regional and vernacular epic and folk traditions have often been seen in opposition to Sanskrit epics and Brahmanical domination,[19] a kind of counterculture that allows for critical expressions of ambivalent and even negative attitudes toward Brahmans and orthodox values.[20] They espouse a holistic 'local world view' or regional integrative process that provides 'an alternative to Brahman elite ideology'.[21] Scholars have thus reflected on the relationship between local folklore and low-caste, tribal, and women's narratives. A. K. Ramanujan classified women's folk tales as a counter-system, an alternative way of looking at the world, which is different from men.[22] While critiquing Ramanujan for seeing the world of women as exclusive, Gloria Goodwin Raheja and Ann Grodzins Gold also argue that women's folk tales often contain everyday forms of resistance that challenge dominance.[23] Verrier Elwin extensively worked on tribal folklore and traditions in central and northeast India, and opened a world of work, love, nature, women, land, poverty, dance, music, and human relations.[24] Hailed as a Dalit thinker,[25] Linda Hess remarks on Kabir's special place in India's oral and expressive culture, where written texts have been turned into a dizzyingly fluid oral performance tradition that offers trenchant critiques of injustices of caste and hypocritical religiosity.[26] Blackburn examines Tamil folklore, particularly bow song stories and performances, and their relationship to Nadars, caste heroes and heroines who died in struggles for social justice, and birth and death stories, which 'transmit that collective identity known as history'.[27] And Eleanor Zelliot talks of how Dalit folklore in Maharashtra is a 'folklore of pride', which centres on ideas that Dalits were creators of culture, 'Lords of the Earth', and militant people, with heroes who used their strength to self-sacrifice for their people.[28] A book on Dalit cultural assertions in Nepal articulates: 'It is about the gods they have worshipped, stories they have told, myths they have created, and beliefs they have held. It is about their music, their songs, and their arts.'[29]

In the north Indian river plains, Dalit folklore has particularly gained attention since the past two decades, as new narratives of Dalit politics have extracted folk heroes, dissenting epic characters, saints and social reformers like Dina–Badri, Deosi and Savari, from folklore. Badri Narayan sees it as subaltern history, where peoples' consciousness, resistance, and political articulation create a folk genre of marginalized sections of society. The educated and politically conscious middle classes of Dalit–Bahujans in Uttar Pradesh and Bihar have played a key role in writing, propagating, and publishing folk literature, and constructing a political language that includes folk symbols as a discursive strategy.[30] For example, folk tales and myths of Dalit women and their role in the 1857 revolt have been effectively used in Uttar Pradesh to challenge dominant historical narratives about the rebellion, as well as to create icons of Dalit feminine and political power. Jhalkari Bai, Uda Devi, and Mahabiri Devi have been projected and hailed as Dalit *virangana*s (brave and heroic women), locked in violent conflict with the British. The production of Dalit folk tales 'represent dissident voices, coexisting with and simultaneously challenging dominant narratives and ideologies'.[31]

In a similar vein, Dina–Bhadri embody a Dalit folk tale, narrated by Dalits from their perspective. The folklore reveals a world in which Dalits carve their own space and question the authority of *savarna* (high) castes. The folk tale has men and women, old and young characters, belonging to both backward and forward castes. It has a low-intensity, everyday presence within Dalit households and villages, but its performance, through tales, ballads, songs, dance, and theatre, is also organized on bigger and special social and cultural occasions. Amateurs and professionals, individuals and groups, mainly from lower castes, contribute significantly towards creating a public folk culture. In different regions and languages of the Gangetic Plains, there have been subtle changes in the characters, incidents, and conclusions of the story. Indrajit Roy notes:

> As with so many folk tales, there is no one plot to which the narrators of this ballad adhere. Not only do the names of the landlords vary, but also their specific indulgences and atrocities as well as the stories of what happened to the two brothers at the end. The apparent absence of a coherent plot is marginal to the symbolic value of the tale, which provides an opportunity for recalling deeds of valour, paying respect to one's ancestors and having some entertainment.[32]

The folk tale of Dina and Bhadri has many versions in Maithili, Magahi, Bhojpuri, Hindi, and Angika languages, spoken in several parts of Bihar, Uttar Pradesh, and Nepali Terai. Various Hindi and Maithili publications of the folk tale describe Dina–Bhadri as caste heroes, with wide popular appeal within the Musahar community.[33] The Maithili folk song narrating the story has been a primary reference point for different sub-regional versions. It was first translated and published in 1885 by George A. Grierson as part of his mammoth 'Linguistic Survey of India'. 'The song of Dina and Bhadri' is placed prominently under the 'Selected Specimens of the Bihari Languages'.[34] It is divided into seven chapters and 410 lines. A summary of the chapters can provide the broad contours of the folktale:

a. The landlord asks Dina and Bhadri to labour in his field. They refuse saying that they live by hunting. Both are beaten and treated shamefully.

b. Dina and Bhadri go out hunting with their uncle Bahoran and are killed by the were-jackal Photra.

c. The spirits of Dina and Bhadri ask the uncle to carry the news of death home. The uncle is afraid to go alone and refuses. There is confusion at home for a week about their non-return. The spirits send word of their death by Ahir Goar.

d. The spirits of Dina and Bhadri disguise themselves as mendicant ascetics and visit their village. After a variety of adventures, they make themselves known to their father and mother.

e. The disguised ascetics depart and carry off Hira Tamolini and Jira Lohaint as their wives.

f. The spirit of Bhadri asks Gulami Jat for milk. He refuses churlishly. Thereupon the two spirits enter the body of were-jackal Photra and overcome Gulami.

g. A Rajput landlord, Jorabar Singh, attacks the marriage procession of the spirits of Dina and Bhadri, and carries it off. Bhadri conquers Jorabar Singh with the help of Gulami Jat.

The folk tale moves in Musahar villages, communities, and castes. It encompasses Dalit and Hindu gods; land, agriculture, forests, rivers, and animals; caste-based rituals and symbols; women and questions of dignity; supernatural beings, both divine and demonic; and violent conflicts between Dina–Bhadri and the Rajput landlord Jorabar Singh, which culminate in the killing of the oppressor. In the folk tale, Dalits worship their gods like Teliya Masan (a kind of ghost or spirit), Dihbar (village god), and Salhes (a god worshipped primarily by Dusadhs), but they also celebrate Hindu festivals like Dussehra and perform extensive Hindu rituals during marriage and cremation. The Dalit folk tale often invokes Hanuman, and Dina–Bhadri frequently mention and worship Ram in desperate situations. People from different low castes like Musahar, Dusadh, Teli, Ahir, and Mali live together, separate from the savarnas, but they also have caste rivalries within themselves. According to the legend, Dina–Bhadri are killed by the were-jackal Photra because the animal is protected from death by the Dusadh God Salhes.

In the folk tale, Dina and Bhadri are experienced hunters in the forests. Other Dalits are agricultural labourers, engaged in rural occupations. However, their economic and social life is eclipsed by caste/village-based discrimination and humiliation that has some unique features in the Gangetic Plains. For example, the prevalence of an oppressive practice of 'obstructing a door',[35] that is, when a landlord wants to compel a Dalit labourer to perform any work, which the latter refuses to do, he sends a male servant to 'obstruct' the door. The servant simply sits on the road and leers constantly at any Dalit woman moving in and out of the house. The family is thus forced into compliance, for the women dare not leave the house, either to fetch water from the well or in the morning for necessary ablutions. In the beginning of the folk tale, when the landlord reaches Dina–Bhadri's home and asks them to work on his land, their mother Nirso says bitterly:

> O Dhami, that you have obstructed my door so early in the morning? You have put your own daughters-in-law and daughters to sleep, and keep them safe in your house, and you have (come here to) see my daughters-in-law and daughters naked and uncovered.[36]

Dina–Bhadri mention the labour system *nafar* or *kamiya* – a local term used for serfs. Landlords give poor labourers advances of grain to support them, and they, along with their children, are bound forever as slaves to the landlord. There are occasional feudal services demanded by the landlords from the tenants. *Savarna* landlords exercise enormous power over the Dalit labourers. However, the folk tale shows that on some occasions, landlords are conciliatory and their power is not absolute. Once the landlord offers higher wages to Dina–Bhadri and when they reject it and refuse to work, the landlord feels that a 'great indignity' has been done to him. The landlord has to tolerate them because the 'enemy' is more powerful: 'If I beat him, wife, today, he will beat me, and therefore, I will not have the courage to do so. Dina and Bhadri have many persons to help them.'[37]

The folk tale has many active and passive women characters, who represent stereotypes while also offering their counter images. In Dina–Bhadri-centred episodes, they appear as mother, daughter, wife, and daughter-in-law, and add their own flavour to village myths and events. Women characters take no independent actions. However, their regular talks, sharing, questions and answers, and critical remarks prepare the grounds to challenge dominant power structures. In fact, it is mainly through a woman's advice that Dina–Bhadri draw a narrative to challenge an age-old repressive practice in the village. Under the dominant social custom, new brides were forced to spend their first night with the *savarna* landlord. It is narrated:

> Dina and Bhadri are taking along their brides' litters and Jorabar Singh stopped them. He asks the litter-bearers, 'where do you come from, where will the litter go? You will not be allowed to go on. Ho, you guards, take off the litter to Kanauli. Who is the man (that dare stop you)? Send him before me.'[38]

The folk tale ends in Kanauli when the Rajput Jorabar Singh is killed in a fight with Dina–Bhadri. The central feature of the folklore is the portrayal of Dina–Bhadri as heroes from below, as voices of the downtrodden. While contending that Indian 'folklore research has been overwhelmingly based on classical, almost exclusively Sanskrit, and largely literary sources',[39] Stuart H. Blackburn contrasts the 'hero' of an aristocratic-courtly society with the local hero in the ballads of Tamil-speaking people, and states:

> As the protector of the lower classes, the local hero necessarily differs from the puranic hero in the content of his acts. While the puranic hero protects the chastity of princesses or does battle for the king, the local hero protects cattle or crops of the village. While the puranic hero challenges forces that threaten to upset the status quo of the kingdom, the local hero opposes casteism and social injustice. Further, the heroism of the local hero is a function of his humanness, not his approximation of a god as with the puranic hero.[40]

Dina–Bhadri, too, are community heroes who embody sacred spirits of ascetics. Elements of the folk tale, comprising their supernatural, magical, and illusionary powers, often use domestic and wild animals such as cow and were-jackal. Dina–Bhadri can cover any distance, transform into different bodies, predict events, and read thoughts. It is thus possible for them to achieve anything for their community. Dina–Bhadri have a natural flair and practice of escaping from death, destruction, and other deadly social traps. They escape from landlords, wild animals, death, drudgery, and other perils, often in a combination.

The folk tale has acquired a variety of offspring, and Musahars have been adding historical and contemporary narratives, songs, local gods, and events to the meta-narrative. Even in the Maithili-speaking regions, there is no single standard format of the folk tale. Some songs and tales have more than ten parts. They are found both in oral and non-verbal forms. In one version, the folk tale narrates in thirteen chapters the hard life of rural people under the local feudal rulers, who are working in collaboration with the main ruler, Jorabar Singh. People are terrified by the brute power of trader Sinuria, village headman Tumri, muscleman Potra, and wicked oculist Nath. The oppressor has more physical and muscle power than Dina–Bhadri, who instead possess divine, spiritual, and dialogical powers. That is why they are ultimately victorious. The folk tale has much to describe about forests, animals, food, clothes, festivals, and customs of both the rulers and Musahars, in a somewhat contrasting and humorous manner.[41]

In Bajjika-speaking[42] regions of Bihar and Uttar Pradesh, the folk tale of Dina–Bhadri has been a living symbol of Musahars' lives and struggles for social justice. In this version, the folk heroes never die and are immortally alive. The brothers are imprisoned by the Rajput landlord after they refuse to work in the field. The landlord tries to burn them alive in a container filled with boiling oil. They fight hard to free themselves, kill the landlord, and ensure freedom for their community. There are three new noticeable aspects here: Dina–Bhadri are placed along with some other folk characters, prominent being Raiya Ranpal who fights with them to defeat the oppressors. There are significant sections describing the admirable strength of Ranpal. The folk tale is not just for Musahars; it also mentions the participation of other low castes like Kumahar (potter), Mali (gardener and flower seller) and Gwala (cattleman), who all remember and pray to Dina–Bhadri. Women folk characters are also recognized and worshipped by Bajjika-speaking Musahars. Kama Mai is a powerful character who can even dry down the mighty river Ganga by her curse. Somaro Dhani is the symbol of agriculture, especially of rice cultivation. It is believed that the negative and positive energies of women figures combine with the endeavours of Dina–Bhadri, and thus should always be invoked.[43]

Amongst the Angika-speaking[44] Musahars, who mostly reside in districts of Bihar and Nepali Terai, the story of Dina–Bhadri is known as a story of Manukhdev on earth, meaning 'God of humans', and its narration has a particular singing style called Ahrayal, in which emotions of heroism, compassion, and fury are evoked through an interplay of instruments and rhythmic dialogues. In the Magahi belt,[45] Dina–Bhadri are *veers*, the braves, whose storylines are not only located in land but also in water and forest areas. Similar to other language versions, land narratives appear when Dina–Bhadri refuse to work as agricultural labourers, but their two main battles against the landlords happen inside the water tanks and forests. This version is entirely in oral form and its contents have been improvized with changing local details, and come to include names of nearby places, rivers, forests, gods, and goddesses.

Maithili-speaking regions of India and Nepal have a rich history of folklore. Believed to be the land of Hindu goddess Sita, consort of Rama, and her father King Janak, the region is also known for the poet Vidyapati and Hindu Vedic sage Yajnavalkya. There are rich collections of Maithili folktales that depict the Hindu religious, mythological, and spiritual values of an agrarian society.[46] The movement for recognition of Maithili as an independent language has been largely dominated by Brahmins, who are now being increasingly interrogated by non-Brahmins.[47] Even some prominent Brahmin Maithili writers have openly asked Maithili organizations to give Dina–Bhadri, Salhesh, and others their due place. At the same time, Dalit folk tales have their own dynamic, and do not easily fit into the prevalent and dominant folkloristic categories. In Nepali Terai, Musahars worship Dina–Bhadri as their *kuldevata* (ancestral deity), *gaundevata* (village deity), and *ghardevata* (home deity).[48]

The Dina–Bhadri folk tale, as a counter-system of a community, remains mostly confined to Dalits, and has an autonomous character that can incorporate changes in its content and form. It demonstrates an active, cross-border, amplifying genre, where a large stock of other folk materials – story, myth, ritual, performance, audience, local character, neighbouring language – have been added. The Dalit community's movement, action, and political consciousness has given a new sense of purpose to recollect and refresh the folk tale in many languages. I will now examine how Dina–Bhadri transform into an important system to accelerate Dalit culture and identity in the everyday life of Musahars, through multiple folk mediums.

Everyday Life and Performance of a Folk Tale

Dina–Bhadri and the performances of their folk tale have a strong identification with the everyday and intimate, domestic and public life of the Musahar community. Ramanujan explains the two categories – *akam* and *puram* (domestic and public) – as useful organizing principles for folklore.[49] Various scholars have examined the relationship between local or low-caste communities, folklore, and its varied performances on the ground. Joyce Flueckiger shows that Chhattisgarh folklore and oral genres are organized not by a performer, a musical instrument, or an autonomous artistic form but by social organizing principles of the community and the people who participate in them.[50] Similarly, according to Velcheru Narayan Rao, folk narratives are transformed by imaginative singers and the changing ideology of the community that deeply identifies itself with the epic and participates in its performances.[51] Susan Wadley discusses *Dhola*, an oral epic about Raja Nal, which is sung and performed mostly by low-caste singers in rural areas of western Uttar Pradesh and eastern Rajasthan, who often 'use their oral traditions to comment on and sometimes to contest the traditional social order, whether the caste system, norms for women, or life itself'.[52] Blackburn's study of Himalayan tribal tales and oral genres of the Apatanis shows how their local stories have been transmitted through time, place, migration, and movement, leaving their mark on expressions of identity.[53] Carola Lorea describes how the widely popular songs of Bhaba Pagla, mostly sung and orally performed by Baul mystic minstrels and fakirs, reveal the role of folklore in cohering a shared identity and heritage among expatriated performers from Bangladesh resettled in West Bengal, many of whom are a displaced community of low-caste practitioners. For them, Bhaba Pagla symbolizes not only a religious guru but also a cultural hero whose songs evoke the nostalgia inevitably associated with migration, and who himself becomes an icon of successful resettlement – with status, popularity, and legacy.[54]

Dalit systems of cohabitation of the Dina–Bhadri folk tale and their community are dynamic, as the performances of the tale have a 'special ability to tell a community's own story and thus help to create and maintain that community's self-identity'.[55] The organization of the Dina–Bhadri folktale within the Dalit community strengthens the core components of folkloric tradition, namely, story, myth, performance, and public. My fieldwork in the Maithili-speaking regions found that the folk tale is increasingly having a greater impact on the everyday life of the community, especially in their public and political spheres. There are places and symbols of Dina–Bhadri, and festivals, plays, and songs on them, which are performed in and around Musahars' villages and are initiated by new Dalit social and cultural groups. People discuss and realize the impact of education, media, migration, and changing social and political situations on the folk tale, which becomes an important catalyst for its creation in Dalit locations.

The divide between folk tale and myth, ritual and symbol, story and song, personal and public, gets blurred in Musahar villages. Dina–Bhadri appear simultaneously in different forms, which also merge with each other. The cultural performances of the Dina–Bhadri folk tale during festivals, religious ceremonies, social events, and political protests modify texts and written words, and are open to interpretations. According to Richard Bauman, such performances of folklore convey a dual sense of artistic action, which involve the performer, the art form, the audiences, and the setting. Instead of a text-centred orientation, the performance becomes 'constitutive of the domain of verbal art as spoken communication'.[56] David Turner describes such cultural performances as 'reciprocal' and 'reflective' where 'the performance is often a critique, direct or veiled, of the social life it grows out of, an evaluation (with lively possibilities of rejection) of the way society handles history'.[57] More recently, examining the vibrant presence of folk performances in the socio-cultural lives of subaltern communities in India, Brahma Prakash sees them as sites of symbolic struggles and contestations.[58] Based on his studies of *bhuiyan puja* (land worship), *bidesia* (theatre of migrant labourers), *Reshma-Chuharmal* (Dalit ballads), and *dugola* (singing duels) from Bihar, and the songs and performances of Gaddar, Jan Natya Mandali, Telangana, he states that 'the centrality of labouring bodies defines the aesthetics and politics of this performance world'.[59] He concludes that such folk performances signify 'a mediation between theatre, social, and cultural process in which social gets theatricalized, culture gets socialized, and aesthetics get politicized'.[60]

Folk theatre is central to the performance of Dina–Bhadri, and has acquired new contours in the past decade among the Musahars. Nandi Bhatia finds popular and regional folk theatre forms like *nautanki, jatra*, and *tamasha* as critical in understanding colonial and post-colonial subaltern voices of resistance, along with mythology, historical stories, oppressed histories, and Anglo-European productions.[61] Scholars have particularly examined Dalit folk theatre and its expressive forms in Maharashtra. Discussing *lavani* and *powada*, Sharmila Rege calls them 'caste-based forms of cultural labour', which have been marginalized by bourgeois forms of art. She sees the roots of these cultural practices in social and material conditions of Dalits and Bahujans, and they continue to relate to the everyday lives, struggles, and labour of different classes, castes, and genders.[62]

In the region of my study, folk theatre based on Dina–Bhadri is village level in nature, and closely connected to contemporary social and political developments. These short-duration plays have evolved the folk tale of Dina–Bhadri into various dramatic forms involving stories, dances, music, and group and individual songs, with an inspiring message in the end. The writers, actors, organizers, and audiences of these plays are mostly Musahars and Dalits. The plays are organized either as part of a wider social event or during festivals and fairs. For example, a popular Maithili play *Dina-Bhadri* consists of five scenes and is of forty minutes' duration. The play was published locally in 2013 and has since been played innumerable times, mainly in Musahar villages. Musahar social and cultural organizations in the region like Musahar Sanskritik Manch, Musahar Youth Sabha, and Lok Shakti Sangathan have enacted this play on stage and streets. The latter also enacted the play in Jhanjharpur during land and water struggles. Chandresh, the writer of the play, explains:

> The storyline of the play is primarily the depiction of the popular story of Dina–Bhadri, when they successfully fought against the atrocities of *savarna* landlords. The dialogues and actions focus on two specific aspects of feudal atrocities – bonded labour and *Dola* system [atrocity on brides], and the courage and protest of the most marginal people. The backdrop of the play has consciously been kept around the areas of Kosi river, where Musahars continue to live under feudal and casteist exploitation. This is not a caste play, but a social play with a message to inspire people to fight for an equitable society. The performances are also socially grounded where people from the downtrodden community act in the play. The play is empowering for the Dalit community.[63]

In Nepali Terai, another Maithili play *Bhaiya Aaile Apan Suraj* (Brother our freedom achieved), written by Rambharosh Kapdi 'Bhramar', was published in 2010. Musahars have enacted this 1-hour play several times, particularly in the Dhanusa district of Nepal in inner Terai. It narrates the lives of Dina–Bhadri as children, young persons, labourers, and warriors, and uses simple, communicative language. According to the writer Rambharosh:

> The original story of Dina–Bhadri must be preserved in the play. However, I have portrayed the folk heroes not as Gods for religious purposes, but more as enlightened, brave, selfless and inspiring human-like characters, who have relevance in the contemporary world. There is much focus on theme and dialogue.[64]

In rural north-eastern Bihar, Dina–Bhadri festivals are usually organized in Musahar villages twice in a year, in March and June, by the agricultural labourers and are public events on public spaces. Plays, dances, songs, and ballads, with music of *mandal* and *dholak* (drumming instruments), are the main features of the festival. A Musahar *dhami* (priest) begins the festival with the worship of Dina–Bhadri, normally by putting two–three new bamboo poles with red and white flags. The idols of Dina–Bhadri are sometimes put at the centre of the cultural performance.[65] There are many Dina–Bhadri temples now in the Gangetic Plains of India and Nepal. Recognizing Dina–Bhadri as important cultural and religious symbols of an ethnic population, the Nepal Academy has published a pictorial book on their origins, places, temples, and festivals.[66] A well-known temple here is in Katiya Khap, in the Satpadi district of Nepali Terai. A fair is held here every year in the month of *Asadh* (August–September) and the ballad of Dina–Bhadri is sung for five nights:

> The songs are in Maithili, but the stories are narrated in a mix of Hindi and Maithili. Three forms are used in the ballad: first is the *marauti* which is used for singing the songs narrating the life and wars of Dina–Bhadri; the second is the *jagar*, which is the narration of their life in the form of a story or *gatha*; and third is the *jhoomar*, which is used at regular intervals to change the mood. *Marauti* and *jagar* have the same content, but there is a difference in their form and style. *Jhoomar*, on the other hand, is sung to [a] fast beat and tells the stories of gods and goddesses, river, mountains, and other natural entities.[67]

Other prominent Dina–Bhadri temples are in Ekausi, Usri Dih, Nawani in Madhubani, and Mangauli Kothi in Samastipur districts. Worships and fairs are organized in most of the temples. In some places, a significant feature is the complete harmony of *dhami*'s feelings (*bhava*) with the characters of the creator. Thus, *dhami* gets possessed by Dina–Bhadri. In Mangauli, the *mridanga* dance is quite popular – an energetic performance by a group of young people, for deflecting the enemy and protecting the folk heroes.[68]

Broadly following the meta-narratives of a folk tale, there are various folk myths of Dina–Bhadri which give concrete details about their birth, place, time, appearance, travel, and family, through the use of oral sources. For example, in the Indo-Gangetic Plains of Nepal, it is believed that they were born in Yogiyanagar of Saptari district of outer Terai, bordering the Sapta Koshi river. Motilal Nepali, a Dalit social and cultural activist researching on Dalit history and culture in Nepal, and belonging to the Dalit Welfare Association at Kathmandu, explains the finer details:

> It is here believed that Dina–Bhadri were brave since their childhood, and in their youth, they had bows and arrows of 80 kilos each, always hanging on their shoulders. Their chests were so solid that even the 84 kilos of boulders did not mean much to them. The story is narrated, sung, symbolized, and celebrated in Yogiyanagar, and the villagers have built a small temple, where the statues of Dina–Bhadri are placed with bows and arrows.[69]

Amidst a cultural and political momentum among the Dalits, the folk myth was also written and published in the Maithili language.[70] Musahars in the

Tirhut region of Bihar believe that the lineage of Dina–Bhadri comes from
Shabari, the *bhil* woman ascetic in the Hindu epic Ramayana. According
to this version of the folk mythology, Shabari lived and died in the forest
zones of Muzaffarpur and Darbhanga districts and her two sons were Dina–
Bhadri.[71] These stories of Dina–Bhadri describing the place of their origin,
time, and situation are also combined with organizing special performances
and celebrations to especially mark the importance of place.

There are folk symbols of Dina–Bhadri, embodying their cultural and
social values. These symbols are visible in Musahar villages, community
places, and homes. The physical characteristics of symbolic objects are simple,
less ritualistic, and carry distinct meanings, which are interpreted differently
in different villages. These symbols are also meant to orient human actions
and behaviours and people are expected to handle them in certain ways.
In the Bajjika-speaking villages of Samastipur district in Bihar, mud scaffolds,
representing Dina and Bhadri, are built outside the homes of Musahars. They
are small, home-made structures without any cover to protect them. According
to villagers, Musahars are identified with mud in north India.[72] Thus their
folk gods are also made of mud, which naturally exists everywhere – in land,
home, water, hand, and body – and the scaffolds are always kept naturally
under the sky. It is believed that Dina–Bhadri wished to be placed outside
the homes under the open sky, so that they can protect Musahar homes and
lives against natural and physical adversities.[73] In some villages of Jhanjharpur
sub-division, there are two black stones outside Musahar homes, symbolizing
the presence of Dina–Bhadri. Villagers explain that their folk gods were living
in deep forests, in dark and dangerous zones, as hunters and wanderers. Black
stones signify the supernatural powers of folk symbols for good and selfless
purposes. They are placed outside the homes to invoke the same spirits, and
to produce positive results for the practitioners. In some households, the black
stones are also placed inside the home, along with the traditional symbols of
the village god Dihbar.[74]

In Saharsa district, two red and white flags hoisted on two long bamboo
poles are visible in the community place (*thaan*) in every Musahar village. Flags
represent the victory of Dina–Bhadri and are also understood as community
assertion of their location. In some villages, there are three flags hoisted, one
being the symbol of the Hindu god Hanuman, who as a devoted companion of
Rama destroyed the kingdom of the devil god Ravana. According to villagers,
the hosting of a third flag signifies that Dina–Bhadri are like Hanuman –
god of victory, supreme destroyer of evil, and protector of devotees.[75]

There are many more folk symbols like the Marauti Sthal (abode of the deceased) – triangle-shaped small stone-carved structures – on spots where the blood of martyrs Dina–Bhadri is supposed to have flown while fighting against the oppressors. The leaves of the Sal tree (*Shorea robusta*) are considered sacred, not to be used for mundane things, because Dina–Bhadri live in the Sal forests.[76]

The folk tale has a living presence among the Musahar community and is constantly changing and evolving. Of course, the story of Dina–Bhadri has deep roots in history, and some of the folklore genres found in the region have ancient precursors. Yet, their evolution and remarkable popularity also signifies contemporary social and political developments in the Dalit public sphere. The performance of the folk tale is mostly a public and collective exercise, where oral transmissions coexist with printed and digital materials. There is a new proliferation of printed and digital texts that have tremendously enhanced the popularity of the tale. The old oral traditions and the new communication materials have encouraged the development of new contents and performances of the folk tale. These developments are most visible today in the arena of environmental movements, which I am dealing with now.

Ecological Ancestors, Saints, and Heroes

> In the west of the village
> A new pond constructed by the landlord sister
> Purain Plant in the Pond
> O Dina! O Bhadri! You will not enter into the pond
> Dangerous snakes hide in the plant
> Let snakes be yoked to chain![77]

Folk literature has rarely been taken as a serious and important source to understand environmental histories, though some work has emerged in recent times. African American eco-literary traditions have particularly touched the subject. Kimberly N. Ruffin, for example, identifies the trend of claiming 'historical figures as ecological ancestors, archetypes who embody honourable actions and attitudes': mythical characters with immense 'physical, botanical and hunting skills', 'expert navigator equipped with not only social and geographic knowledge but also the spiritual fortitude to liberate herself and others from environments made unforgiving by oppressive human systems',[78] or deep spiritualists committed to humans and nonhuman nature.

In the context of India, examining the history of pre-modern tanks in Karnataka through oral narratives like folk stories, songs, and legends, Esha Shah states that they show how tank technologies 'were socially embedded in societies and economies that were organized for warfare, sustained sharp social hierarchies, and were often violent to women and people from lower castes'.[79] The pioneering work of Komal Kothari on folklore of Rajasthan underscores that musical and performance traditions (puppetry, folk songs, and stories) of low castes are embedded in land, water, agriculture, irrigation, and livestock. Dalit folklore is filled with tales of their relationship with the environment, along with their ecological interactions with dominant castes. In the process, they construct ecological mystiques, while also attempting to divest the authority and sanctity built around dominant gods and castes. According to Kothari, through oral genealogists, they narrate their past to reaffirm the present.[80] Anthropologist Anand Pandian explores diverse practices of cultivation among the Piramalai Kallar caste, condemned as a 'criminal tribe' for long, in Tamil Nadu, and their use of material labour and environment to carve ethical lives.[81] Dhangars, low-caste pastoralists of Maharashtra, sing *ovi*s – oral stories of their gods – which are often about 'divine figures who take birth as humans, do battle with demons, and eventually succeed'.[82] In my earlier work, I, too, have examined some of the historical, mythical, and contemporary trends of Dalit activism and thought on environment, and gleaned the emergence of 'Dalit eco-literature' – poems, paintings, stories, music and folklore – coming from diverse regions and communities:

> Dalit myths and legends of Mayabel and Jasma, Deena and Bhadri underline their dreams and desires for ecological belongings, against their suffering, sacrifice, and alienation. They have their gods and goddesses, pujas, and festivals – for example, Kattamaisamma (discoverer of the tank system and goddess of water), Potaraju (protector of soil and fields), Yandi (marvel of technological knowledge), Nuakhai, Dalkhai, Duma, and Maati Devi – to celebrate and highlight their ecological capacity and connectivity to natural elements against all odds.[83]

Musahars have a treasure of folksongs and folk tales, which 'are ritualised reproduction of Musahar suffering as well as their acumen for survival'.[84] They are living cultural resources for the community, especially for women and youth, and strengthen collective identity and community solidarity.[85] Narratives of Dina–Bhadri, with their several versions, are an integral part of folk culture of the community. The folk tale and its cultural performance are deeply entwined with land, water, and forest, and lives of different castes who live on them. It is not simply that environment creates a stage on which Musahars live and Dina–Bhadri act. Rather, the environment is an integral part of their culture, which in turn expresses how Dalits engage with their natural and social worlds. Musahar culture has a distinctive relationship with its environment, which is different from that with the *savarna*s. Dalits give their nature relationship a unique meaning and create a cultural landscape of their own.[86] Dina–Bhadri appear as wise ancestors, well versed with nature, earth, environment, life, and death. People believe that their ancestors have given a natural and social relevance to the landscape, instilling it with their humanness and heroic strength. They are like saints, leading a selfless life, and their heroic deeds liberate the earth from caste bondages.

Musahars have been known in the region for their traditional skills of measuring and assessing the quality of soil. They carry out some of the hardest agricultural activities of soil-cutting, earth-removing, and heavy spade work on hard soils, and their rooting in land is strong. It is commonly believed that Musahars do not need an inch tape, or a modern measuring device, or the advice of an engineer, to measure the quality of the soil or suitability of the land. There are a large number of dams, ponds, bridges, and multi-storeyed buildings where their labour and skills have been applied for assessing, digging, or measuring the soil. When I visited Kohlara village in Samastipur district, it was the agricultural season for paddy cultivation. Musahar agricultural labourers working in the fields narrated to me a folk song which expresses their gratitude to Dina–Bhadri for their deep knowledge of land, soil, and weather, which has been passed to their living generations. After a day-long work, they normally sing this song in evenings in groups at a community place:

> *Emerged from the soil*
> *Played into the soil*
> *Made up of soil*
> *Ended in soil*
> *Oh Dina-Bhadri! Oh our forefathers!*
> *We know today the heart of soil and land everywhere because of you.*[87]

Musahar folk plays and songs have rich accounts of land, agriculture, weather, crops, domestic animals, and food, which encompass the close association of Dina–Bhadri with rural landscapes. They characterize their relationship with land and agriculture differently from that of *savarna*s. Unlike the *savarna*s, Musahars have no claim over land, but they have a share in the crops, and thus they celebrate the latter. Dina–Bhadri are eulogized and remembered for making people physically and mentally strong to enable cultivation for the whole year. In a rice-cultivation song, images of Dina–Bhadri are mirrored in various processes of human labour – preparing the rice, husking, cleaning, and collection. Like saint saviours and creators, Dina–Bhadri are deeply involved in the pursuit, along with humans.[88]

Rivers define the Gangetic Plains and its population. Fishing in rivers and ponds has been very important for Dalit communities, strengthening their material and nutritional conditions and livelihoods, bringing them additional food and money, and giving them time and space to build communal solidarities. In the cultural performances of Dina–Bhadri, there are energetic water narratives. At times, they bless the Musahars at the critical moment of digging a pond or they free the pond from a curse of poisonous snakes. They impart the people with wisdom to make peace and partnership with mighty rivers and at times, the heroism of Dina–Bhadri is in full gaze in fighting the river robbers.

In the Bajjika-speaking Musahar villages of Samastipur district, I came across a river folksong, which offers an account of the special bond between Dina–Bhadri and the Kamla river. Musahar women often sing it in a group during the monsoon season. The song describes the marriage of the Kamla river and her anxious wait for the timely arrival of Dina–Bhadri for a peaceful and joyous ceremony. Dina–Bhadri, meanwhile, are busy making every effort to arrange things required for an auspicious wedding. I was told that this folksong is also performed with instruments in village festivals by Musahar cultural groups:

> Sitting on a loft, River Kamla is looking outside from her windows,
> Waiting for Dina–Bhadri, she is thinking about their whereabouts.
> Dina–Bhadri are away, thinking how to make a palanquin,
> And where to get the woods, for the marriage of Kamla.[89]

Dina–Bhadri folk tales and cultural performances have received a new lease of life since the 1990s, after the emergence of Dalit organizations and struggles over land and water rights in the region. In fact, public performances of Dina–Bhadri tales and their festivals became a contentious issue with the local elites from that time, as Musahars were claiming public spaces hitherto dominated by them.

According to historian Kumar Suresh Singh, expansion of capitalist agrarian relations and continuing population pressures helped to dissolve the traditional patron–client relationships. With this background, there arose stirrings of discontent among the Musahar sharecroppers and labourers in Bihar, which led to autonomous protest movements under the leadership and organization of militants drawn from their own community. Musahars have had 'a new tide of consciousness' and have fought glorious struggles 'for their rights, identity and self-respect'.[90] Conflicts over land distribution, livelihood, wages, displacement, and rehabilitation between the landowning *savarna* caste and the landless Dalit agricultural labourers have become widespread and frequent in the Gangetic Plains. Musahars have been mobilized over issues of iniquitous land holdings, money lending, and social injustices. The folk tales of Dina–Bhadri, along with idols, flags, songs, and dances, have been used in the course of claiming access and ownership over private, ceiling, surplus, or public common land. In Madhubani, Darbhanga, and Samastipur districts, where Musahars have been struggling for the payment of minimum agricultural wages and the allocation of land rights over surplus and common lands, songs and plays of Dina–Bhadri are performed in the villages regularly by their own cultural and social groups. Performers, mostly Musahar youth, use costumes and musical instruments to create an eventful atmosphere, and narrate the story of Dina–Bhadri, with a contemporary emphasis on land struggles. As has been remarked:

> These ample living cultural resources consolidate processes of collective identity formation and strengthens community ties. These communitarian ties help the Musahars withstand onslaughts on their precarious status, invent and deploy coping strategies for survival, nurture collective representation by inventing new cultural idioms that strengthen processes of social awakening and questioning the dominant social order and active participation in democracy.[91]

It is a common practice by Musahars to place long bamboo poles with flags, symbolizing Dina–Bhadri, on the captured land.[92] The original folk dialogues between Dina–Bhadri and the landowner over the quantum of agricultural wages have been subtly modified to include contemporary land issues, mainly in popular songs and plays.[93] Musahar agricultural labourers thus offer 'poly-vocal' and 'nuanced perspectives that foregrounded their concerns with social equality'.[94]

Since 2000, Musahars have initiated a movement to claim access and ownership of the village ponds and have also undertaken community conservation works to rejuvenate them. There are numerous ponds in the region, criss-crossed by rivers and streams, giving the impression of a place that almost floats on water.[95] The ponds are interconnected amongst themselves and also with the river, resulting in their smooth recharging through floodwater.[96] According to one report, in Madhubani district alone, there are more than 1,500 small and big ponds. As this region is affected by floods every year, floodwater collects in the ponds, and rivulets are available in abundance for fishing almost throughout the year.[97] Further, shifts in the course of the Kosi, Kamala, and Bagmati rivers are a recurring feature, which convert different patches of the area into water bodies full of fish. The region has supplied fish to the whole of Bihar and even to outside markets in West Bengal and Assam. However, most of the ponds, traditional or government made, have been under the control of *savarna* landlords and moneylenders.

While asserting their claims over ponds, Musahars have attempted to evolve diverse mediums to strengthen processes of social questioning and have sometimes offered parallel structures to the past and present cultural order. At several disputed pond points, one sees the presence of symbols of Dina–Bhadri. These symbols have not only provided strength to Musahars' lives and struggles but they have also shaped their local cultures and contexts, as well as left their imprints on changing locations and understandings of water in the Mithila Dalit society. In one of their meetings, organized in the Sirpur Musahari village, Musahars sat under two flags, hoisted on two long bamboos. Asharfi Sadai narrates:

> Every year in June, we remember Dina–Bhadri. In small groups, we go from house-to-house and village-to-village to collect rice or paddy, so that their memory is kept alive. They are remembered and worshipped because they are our true saviours, who suffered and sacrificed for our struggles over natural resources. They are our only gods. Their sacrifice is ours.[98]

Musahars' myths and legends, invoked in the course of their water struggles, are inundated with violence, war, killing, death, and harassment. Their ecological landscapes are marked by sacrifices and deaths, incurred in order to defend the water rights of their people. Their fight for water, land, and labour throws up ecologies of suffering. At the same time, it also conjures myths as ecological creations. Their suffering is juxtaposed to celebration and construction of their heroes, which symbolize the ecological visions of Dalits. In Sirpur Musahari, the story of Dina–Bhadri appears along with a goddess, pond, labourer, and peasant. Ravi Sadai narrates:

> Dina and Bhadri saw a dream of Goddess Bageshwari, who asked them to go to Lari Larwar, where Musahars were digging a pond, but were decimated by two powerful and cunning persons, Hansraj and Bansraj, who wanted the Musahar

labourers to be buried underground. The goddess ordered Dina–Bhadri to go to Lari Larwar as peasants and to free the oppressed Musahars. Finally, Dina–Bhadri reached the place and rescued the labourers by killing Hansraj and Bansraj.[99]

In Sohrai Brahamotor village, Dina–Bhadri are called the *bir purush* (brave men), who were born in the north-east part of Madhubani district. Hari, a local activist of Lok Shakti Sangathan, retells another colourful tale, which according to him is sung and enacted by Musahars in their festivals and struggles:

> Dina was the younger one, and their parents were Kalu and Nirsaun. They became fighters and warriors in their early childhood. By the age of 12, they had fought many battles to protect the poor labourers from the exploitation of rich landlords. Their first major struggle was against Dhamiya Kanaksingh, who was the stick wielder of the ruling landlord of Ruchauli kingdom. They fought hard against a contractor Kangaliya Dusadh, who had made hundreds of Musahars bonded labourers. They fought for forests and rivers.[100]

Musahar villages and households repeatedly keep their symbols alive in the course of their water struggles. One constantly comes across a small mud-carved raised platform. This represents the Dina–Bhadri *asthan* (place), a point from where Musahars begin their water journeys and sustain their ecological struggles. Americi Devi, who has participated in land and water struggles, narrates:

> Dina–Bhadri follow us wherever and whenever we go for our water. They stand with us in our struggles over ponds and land. They give us strength, and encourage us to never give up. They are our ecological inspirations and our ecological ancestors.[101]

Deepak Bharti, an activist of Lok Shakti Sangathan, an independent, non-party organization of Musahars and Dalits, which has been in the forefront of water movements, describes the broader horizons of the struggle:

> A region which has a rich history of ponds and tanks, and diverse ways of water harvesting and harnessing its fruits, witnessed a cry for justice from Dalits, stating that these water bodies should be restored to them. Mallahs and Musahars, along with their organization, were not only willing to revive the ponds and the streams but also wanted to regain rights over them. The process of recovery also addressed people's cultural and spiritual needs to connect with their physical and natural environment in intimate ways.[102]

Along with invoking Dina–Bhadri, Musahars lament the burdens of dominant Hindu mythologies. Brahmin-washed images of the Hindu religion are pervasive in the region and have created almost iron-clad legends, which marginalize Dalit histories and heroes. These powerful images have also been a part of conspicuous ecological sensibilities in the region. However, by celebrating Dina–Bhadri, Musahars create fissures in these dominant images. Dina–Bhadri also help in erasing the ecological burdens of Hindu Tridevs (three supreme gods: Brahma, Vishnu, and Shiva). They put forward the violent struggles and sacrifices made against oppressive human systems and establish ecological entitlements for Dalits.

The relevance of folklore in Indian society, and its spread in local–regional, rural–urban, elite–subaltern, and men–women contexts has been widely analysed. Folk genres have been conceptualized into different categories like its 'classical' and 'modern' origins, 'great' and 'little' traditions, and 'domestic' and 'public' arenas. Even while there have been some folkloristic studies of Dalit, tribal, and other lower-caste communities, the story of Dina–Bhadri, and especially its relation to a Dalit eco-politics, has been understudied within the field. A study of Dalit folklore of Dina–Bhadri sheds light on its distinct specificities and its intricate relationship to caste, community, Dalits, and environmental politics.

Dalits create rich tales and stories on varied aspects of their lives, which evolve from their experiences, which are often in a complex relationship with dominant cultures and symbols. Their natural, supernatural, physical, mythical, magical, religious, and cultural imaginations migrate between two different worlds – of Dalits and *savarna*s – for feeding, nesting, and flying. Their most popular versions are from the 'ground up', where folk tales originate from actual life situations and gradually evolve to display festivity, collectivity, celebration, competition, resistance, sacrifice, and liberation. As the folk tales travel over years and spaces, and evolve according to varied environments, they develop new narratives, and acquire multiple forms and performances. The energetic narration of Dina–Bhadri in the Indo-Gangetic Plains of India and Nepal demonstrates the potential of folk narratives, co-existing with myriad initiatives of social–political organizations of Dalits. It also explains how folklore traditions continue their momentum with increasing mobilization and movements of low-caste people across the borders. The resources for such mobilizations are not invented suddenly in the 2000s; rather, they are based on, and draw from, a rich repertoire of cultures and memories.

Folk tales are visible in Dalits' everyday cultural, public, and political spaces, and are place- and time-tested. In the recent past, community outreach of folk genres has been spectacular. Musahars migrated to several places in the last century and Dina–Bhadri tales have reached new villages, regions, and countries. Spoken and written, oral and visual forms have developed congenial relations between them, where old and new communication forms have intersected to play important roles in spreading the folk tale. There are no neat structures, as the forms and contents of the folk tale are cluttered; yet they narrate a story of community and conflict. Largely low castes are present and participate in the cultural performance of the folk tale of Dina–Bhadri, as they perceive it as a narrative of them, by them, for them, and from them. In the culturally vibrant Mithila region, Dina–Bhadri and Dalit folk genres mostly stay outside the rich literary tradition of Maithili language. In spite of the growth of Dalit public spheres in north India, a divide exists on the ground between 'Indian' and Dalit folklore systems.

The folk tale of Dina–Bhadri is now also about natural resources and local movements. While rooted in a distant past, it is deeply connected to the contemporary and the seasonal. It is situated in the present struggles of Musahars over lands, ponds, and rivers, particularly in the agricultural seasons. The running memories of the past and the pressing needs of the present converge creatively to expand the Dina–Bhadri story. The North Indian River Plains are characterized by heightened environmental risks and conflicts, and Dalits are particularly vulnerable to multiple pressures. Though the cultural landscape of Mithila is inundated with dominant Hindu religious symbols and motifs, Dalits are increasingly asserting their environmental rights through their own motifs. The environmental symbolism of a folk tale provides the Musahars with a cultural and political tool box through which they connect their past, present, and future, and assert their ecological agency and stewardship. Dina–Bhadri appear in the environmental field as ones who are deprived of natural resources, but who bravely fight and move beyond their oppression. The Dina–Bhadri narrative tradition provides a folkloric reception of a 'Dalit past', but its ecological performance is a project of the Dalit present and future.

In the upcoming essay, my primary focus is on specific subjects that encapsulate the ecological insights of Dalits: rivers, commons, place, and the earth. Drawing from Dalits' eco-literary traditions, I elucidate how these key aspects of the environment become central to their identity formation and contribute to the establishment of community collectives and cultural assertions. The Malo fisherfolk, residing along the banks of the river Titash, undertake transformative journeys towards freedom and affirmation, which hold significance not only for the Dalit author but also for the Malo community.

Notes

1. Antonio Gramsci, *Selections from the Prison Notebooks* (London: Lawrence and Wishart, 1971).
2. Mikhail Bakhtin, *The Dialogic Imagination*, trans. Caryl Emerson and Michael Holquist (Austin: University of Texas Press, 1981).
3. Stuart Hall, 'Encoding/Decoding', in *Culture, Media, Language: Working Papers in Cultural Studies*, ed. Stuart Hall, D. Hobson, A. Lowe, and P. Willis (London: Hutchinson, 1980).
4. Raymond Williams, *The Sociology of Culture* (Chicago: University of Chicago Press, 1981), 203–05.
5. Sanjay Kumar, Arvind Mishra, Badri Narayan, and Rafiul Ahmed, 'Representation, Resistance, and Identity: The Musahars of Middle Gangetic Plain', in *Interrogating Development: Insights from the Margin*, ed. Frederique Apffel-Marglin, Sanjay Kumar, and Arvind Mishra (New Delhi: Oxford University Press, 2010), 153.
6. Badri Narayan, 'Myth, Culture, and Democracy', in *The Marginalized Self: Tales of Resistance of a Community*, ed. Rahul Ghai, Arvind K. Mishra, and Sanjay Kumar (Delhi: Primus Books, 2020), 72.
7. For details on Musahars of Nepal, see Lynn Bennett, Dilli Ram Dahal, and Pav Govindasamy, *Caste, Ethnic and Regional Identity in Nepal: Further Analysis of the 2006 Nepal Demographic and Health Survey* (Maryland: Macro International Inc., 2008).
8. *Census of India 2011* (New Delhi: Government of India, 2011), www.censusinida.gov.in, accessed 20 January 2024.
9. *National Population and Housing Census 2011* (Kathmandu: Central Bureau of Statistics, Government of Nepal, 2012).
10. C. K. Lal, 'Cultural Flows across a Blurred Boundary', *South Asian Himal*, 1 February 2002, https://www.himalmag.com/cultural-flows-across-a-blurred-boundary/, accessed 20 January 2024.
11. Anil Agarwal and Ajit Chak, eds., *State of India's Environment, A Citizen's Report: Floods, Flood Plains and Environmental Myths* (Delhi: Center for Science and Environment, 1991).
12. Bhim Subha, *Himalayan Waters: Promise and Potential, Problems and Politics* (Kathmandu: PANOS South Asia, 2001); Nanda R. Shrestha and Dennis Conway, 'Ecopolitical Battles at the Terai Frontier of Nepal: An Emerging Human and Environmental Crisis', *International Journal of Population Geography* 2, no. 4 (1996): 313–31.
13. For further details on Musahars' struggles over land and social issues in Bihar, see Mukul Sharma, 'The Untouchable Present: Everyday Life of

Musahars in North Bihar', in *Village Society*, ed. Surinder S. Jodhka (New Delhi: Orient BlackSwan, 2012); Mukul Sharma, 'Deena–Bhadri's Sacrifice Is Ours: Everyday Life of Musahars in North Bihar', *Labour File* 4, nos. 5–6 (May–June 1998): 3–25.

14. Alan Dundes, *The Study of Folklore* (London: Prentice-Hall International, INC, 1965).

15. Vinay Dharwadker, ed., *The Collected Essays of A. K. Ramanujan* (New Delhi: Oxford University Press, 2006).

16. Jawaharlal Handoo, *Folklore: An Introduction* (Mysore: Central Institute of Indian Languages, 1989).

17. Alf Hiltebeitel, *Rethinking India's Oral and Classical Epics: Draupadi among Rajputs, Muslims, and Dalits* (Chicago: The University of Chicago Press, 1999).

18. Arjun Appadurai, F. J. Korom, and M. A. Mills, 'Introduction', in *Gender, Genre, and Power in South Asian Expressive Traditions*, ed. A. Appadurai, Frank J. Korom, and Margaret A. Mills (Philadelphia: University of Pennsylvania Press, 1991).

19. John D. Smith, *The Epic of Pabuji: A Study, Transcription and Translation* (Cambridge: Cambridge University Press, 1991).

20. Brenda E. F. Beck, *The Three Twins: The Telling of a South-Indian Folk Epic* (Bloomington: Indiana University Press, 1982).

21. Gene H. Roghair, *The Epic of Palnadu: A Study and Translation of Palnti Virula Katha, a Telugu Oral Tradition from Andhra Pradesh, India* (Oxford: Clarendon Press, 1982).

22. Dharwadker, *The Collected Essays of A. K. Ramanujan*.

23. Gloria Goodwin Raheja and Ann Grodzins Gold, *Listen to the Heron's Words* (Berkeley: University of California Press, 1994).

24. For details on Elwin, see Ramachandra Guha, *Savaging the Civilized: Verrier Elwin, His Tribals, and India* (UK: Penguin Books, 2014).

25. Milind Wakankar, *Subalternity and Religion: The Prehistory of Dalit Empowerment in South Asia* (New York: Routledge, 2010).

26. Linda Hess, *Bodies of Song: Kabir Oral Traditions and Performative Worlds in North India* (New York: Oxford University Press, 2015).

27. Stuart H. Blackburn, *Singing of Birth and Death: Texts in Performance* (Philadelphia: University of Pennsylvania Press, 1988).

28. Eleanor Zelliot, *From Untouchable to Dalit: Essays on the Ambedkar Movement* (New Delhi: Manohar, 1992), 318.

29. Diwas Raja Kc, ed., *Dalit: A Quest for Dignity* (Kathmandu: Nepal Picture Library, 2018), 114.

30. Badri Narayan, *The Making of the Dalit Public in North India, Uttar Pradesh, 1950–Present* (New Delhi: Oxford University Press, 2011), 97–122.

31. Charu Gupta, *The Gender of Caste: Representing Dalits in Print* (Ranikhet: Permanent Black, 2016), 109–10.

32. Indrajit Roy, 'Utopia in Crisis? Subaltern Imaginations in Contemporary Bihar', *Journal of Contemporary Asia* 45, no. 4 (2015): 648.

33. A few publications in Hindi and Maithili are Prafulla Kumar Singh 'Maun' and Ashwini Kumar Alok, eds., *Dina–Bhadri: Musaharon ki Samagra Sanskriti* (New Delhi: Samyak Prakashan, 2015); Mahendra Narayan Ram, *Panchalok Devta* (Patna: Bihar Hindi Granth Academy, 2003); Virendra Kumar Singh, *Hamare Lok Devi-Devta* (Patna: Samiksha, 1999); Shiv Prasad Yadav, ed., *Maithili Dalit Lokgatha ao Sanskriti* (Delhi: Sahitya Academy, 2015).

34. G. A. Grierson, 'Selected Specimens of the Bihārī Language', *Zeitschrift der Deutschen Morgenländischen Gesellschaft* 39, no. 4 (1885): 617–73.

35. Grierson, 'Selected Specimens of the Bihārī Language', 656.

36. Ibid., 656–57.

37. Ibid., 658.

38. Ibid., 671.

39. Stuart H. Blackburn, 'The Folk Hero and Class Interests in Tamil Heroic Ballads', *Asian Folklore Studies* 37, no. 1 (1978): 133.

40. Ibid., 134.

41. M. N. Ram and F. Paswan, eds., *Dina–Bhadri Lokgatha* (Delhi: Sahitya Academy, 2012).

42. Bajjika has been classified as a dialect of Maithili language.

43. Telephonic interview with Ashwini Kumar Alok, 15 May 2020. Alok, a writer and journalist, has published articles on the Dina–Bhadri folk tale. Also see 'Maun' and Alok, *Dina–Bhadri*.

44. Angika is closely related with the Maithili language. It is spoken in the Bhagalpur, Munger, Purnia, and Banka districts of Bihar and the Morang district of Nepali Terai.

45. The Magahi belt is derived from ancient kingdom of Magadha – areas south of the river Ganga in Bihar. A Magahi-speaking population is found in nine districts of Bihar and seven districts of Jharkhand.

46. Ram Dayal Rakesh, *Folk Tales from Mithila* (New Delhi: Nirala Publications, 1996).

47. Mithilesh Kumar Jha, *Language Politics and Public Sphere in North India: Making of the Maithili Movement* (New Delhi: Oxford University Press, 2018).

48. Bhabani Pokhrel, 'Strained Identity: Cultural and Religious Rituals of a Musahar Community', *Social Inquiry* 2, no. 1 (2020): 128–50.

49. Dharwadker, *The Collected Essays of A. K. Ramanujan*, 41–75.

50. Joyce Burkhalter Flueckiger, *Gender and Genre in the Folklore of Middle India* (Itacha: Cornell University Press, 1996).

51. Velcheru Narayana Rao, *Text and Tradition in South India* (Albany: SUNY Press, 2016), 304.

52. Susan Snow Wadley, *Raja Nal and the Goddess: The North Indian Epic Dhola in Performance* (Bloomington: Indiana University Press, 2004), 4.

53. Stuart H. Blackburn, *Himalayan Tribal Tales: Oral Tradition and Culture in The Apatani Valley* (Leiden: BRILL, 2008), 4.

54. Carola Erika Lorea, *Folklore, Religion and the Songs of a Bengali Madman: A Journey between Performance and the Politics of Cultural Representation* (Leiden: BRILL, 1987).

55. Stuart H. Blackburn and Joyce B. Flueckiger, 'Introduction', in *Oral Epics in India*, ed. Stuart H. Blackburn, Peter J. Claus, Joyce B. Flueckiger, and Susan S. Wadley (Berkeley: University of California Press, 1989), 11.

56. Richard Bauman, *Verbal Art as Performance* (Rowley: Newbury House, 1977), 11.

57. Victor Turner, *The Anthropology of Performance* (New York: PAJ Publications, 1988), 22.

58. Brahma Prakash, *Cultural Labour: Conceptualizing the 'Folk Performance' in India* (New Delhi: Oxford University Press, 2019), 57.

59. Ibid., xiv.

60. Ibid., 283–04.

61. Nandi Bhatia, *Acts of Authority/Acts of Resistance: Theatre and Politics in Colonial and Postcolonial India* (Ann Arbor: The University of Michigan Press, 2004), 8.

62. Sharmila Rege, 'Conceptualising Popular Culture: "Lavani" and "Powada" in Maharashtra', *Economic and Political Weekly* 37, no. 11 (2002): 1038–47.

63. Telephone Interview with Chandresh, 10 May 2020.

64. Telephone Interview with Rambharosh Kapdi 'Bhramar', 2 June 2020.

65. Visit to Andhra Musahari village, Madhubani District, June 2017.

66. Nepal Academy, *Dina–Bhadri: Sachitra Sankalan* (Kathmandu, 2014).

67. Narayan, 'Myth, Culture, and Democracy', 73.

68. Visit to Bakhtiarpur village, Samastipur District, June 2017.

69. Interview with Moti Lal Nepali, September 2016.

70. Ram, *Panchalok Devta*.

71. Visit to Chajjan-Gangaram village, Muzaffarpur District, June 2017.

72. In north Bihar, mud has complex meanings, with distinct gender and class dimensions, and social and political identities. Mud naturalizes discrimination at the origin of 'dirt'. Historical and political circumstances suggest that mud is not dirt; it becomes 'dirt' when other kinds of dirt lose their meaning. See Luisa Cortesi, 'The Muddy Semiotics of Mud', *Journal of Political Ecology* 25, no. 1 (2018).

73. Visit to Bakunia Bichali village, Saharsa District, June 2017.

74. Visit to Daldal Musahari village, Madhubani District, June 2017.

75. Visit to Barhara village, Saharsa District, June 2017.

76. Visit to Kubaul Musahari village, Darbhanga District, June 2017.

77. Song narrated during visit to Moghlaha village, Madhubani District, June 2017.

78. Kimberly N. Ruffin, *Black on Earth: African American Ecoliterary Traditions* (Athens: The University of Georgia Press, 2010), 85.

79. Esha Shah, 'Telling Otherwise: A Historical Anthropology of Tank Irrigation Technology in South India', *Technology and Culture* 49, no. 3 (2008): 673.

80. Rustom Bharucha, *Rajasthan, an Oral History: Conversations with Komal Kothari* (Delhi: Penguin, 2003), 32–33.

81. Anand Pandian, *Crooked Stalks: Cultivating Virtue in South India* (Durham: Duke University Press, 2009).

82. Anne Feldhaus, Ramdas Atkar, and Rajaram Zagade, eds. and trans., *Say to the Sun, 'Don't Rise,' and to the Moon, 'Don't Set': Two Oral Narratives from the Countryside of Maharashtra* (New York: Oxford University Press, 2014), 16.

83. Mukul Sharma, *Caste and Nature: Dalits and Indian Environmental Politics* (Delhi: Oxford University Press, 2017), xxvi.

84. Arun Kumar, 'Culture, Development, and Capital of Farce', in *The Marginalized Self*, ed. Ghai, Mishra, and Kumar, 45.

85. Rahul Ghai, Arvind K. Mishra, and Sanjay Kumar, 'Introduction', in *The Marginalized Self*, ed. Ghai, Mishra, and Kumar.

86. My early essay has highlighted that Dalit eco-experiences have their own vibrancy and dynamism. Living with nature, they are constantly negotiating with, and challenging, caste domination, while simultaneously articulating their environmental imagination. Dalit thinkers and contemporary excavations of Dalit memory create varied and alternative spatial and social metaphors around environment.

87. Visit to Kohlara village, Samastipur District, June 2017.

88. Visit to Fatki Musahari village, Madhubani District, June 2017.

89. Visit to Morwa village, Samastipur District, June 2017.

90. K. S. Singh, 'Musahar: Community, Context and Equality', in *Asserting Voices: Changing Culture, Identity and Livelihood of the Musahars in the Gangetic Plains,* ed. Hemant Joshi and Sanjay Kumar (Delhi: Deshkal Publication, 2002), 142.

91. Ghai, Mishra, and Kumar, 'Introduction', 11.

92. Sharma, 'Deena–Bhadri's Sacrifice Is Ours'.

93. Visit to Kothu Uttarwari village, Madhubani District, June 2017.

94. Indrajit Roy, 'Emancipation as Social Equality: Subaltern Politics in Contemporary India', *Focaal* 76 (Winter 2016): 17.

95. Coralynn V. Davis, 'Pond-Women Revelations: The Subaltern Registers in Maithil Women's Expressive Forms', *Journal of American Folklore* 121, no. 481 (2008); Sharadini Rath, 'A Journey through Madhubani' (2015), https://www.phalanx.in/pdf/dini.pdf, accessed on 12 February 2015.

96. V. Jha, 'Sustainable Management of Biotic Resources in the Wetlands of North Bihar, India', in *Aquatic: Conservation, Restoration and Management,* ed. T. V. Ramchandra, N. Ahalya, and C. R. Murthy (New Delhi: Capital Publications, 2005).

97. Frances Sinha, K. A. Srinivasan, Rajiv Kumar Singh, and Viji Srinivasan, *The Blue Revolution: Case-Study of Women in the Inland Fisheries Sector* (Delhi: Har-Anand Publications, 1994).

98. Interview with Asharfi Sadai, April 2012.

99. Interview with Ravi Sadai, April 2012.

100. Interview with Hari, April 2012.

101. Interview with Americi Devi, April 2012.

102. Interview with Deepak Bharti, April 2012.

4

Rivers and Commons

Titash and Malos

A river has its philosophic aspect, not only an artistic aspect. Like time, it flows on endlessly. Time in its ceaseless course is witness to events as they take place and subside, and to human demise. So many lives have ended in horrible deaths – from starvation, suicide, or another's evil deed. And then, again, so many lives are born through time, unmindful of the hundreds of deaths around. Titash, too, flowing along its course, has heard many cries of grief at the death of dear ones and has felt the tears of the grieving mingle with its waters.[1]

Commons, including rivers, water bodies, forests, land, air, and the earth, have served as a crucial reference point for various forms of environmentalism. However, when viewed from a Dalit perspective, commons acquire a multi-layered and intricate significance that encompasses social, economic, and environmental dimensions. They encapsulate the inherent tensions and conflicts between social integration and cultural diversity, playing a pivotal role in the expression and resolution of socio-spatial transformations within society. Moreover, commons serve as sites of Dalit creativity, enabling the exploration of self, space, spirituality, and freedom. Through the narratives, metaphors, and collective memories associated with commons, Dalit writers construct a historical account deeply rooted in the earth itself.

Dalit writer Adwaita Mallabarman was a poor Malo, born in 1914 in Gokanghat village, beside the river Titash, near Brahmanbaria town in Comilla district of present-day Bangladesh (it was East Bengal in undivided India until 1947, then East Pakistan until 1970). He lost his parents at an early age and lived with his uncle in the village until his teenage years. The Malo community raised subscriptions to support him, and he was the first child in the village and the nearby area to finish school. Mallabarman could not continue with college education because of financial problems and migrated to Calcutta (now Kolkata) in search of work. In 1950, he completed *Titash Ekti*

Nadir Naam in the Bengali language, and after a few months, he died from tuberculosis in 1951 at the young age of thirty-seven. *Titash* was published in Kolkata five years after his death and became a highly acclaimed novel of Bengali literature. It was also made into a film by Ritwik Ghatak and a play by Utpal Dutta, preeminent Indian film and playwright-directors.

Titash has been richly referred to and reviewed as individual and collective journeys of Malo youth, women, children, and old – in their quest for social awakening, belonging, knowledge, and education – which were crushed by natural catastrophe, modernization, and sectarian conflicts;[2] as the slow and painful deterioration of places, communities, and personal relationships, where even the mighty Titash river began to behave strangely, as if in response to the general disintegration of all things;[3] as Dalits being reduced to a 'surplus population' by the larger forces, pushed to the periphery and the outside, where the subjectivity of 'broken men' is determined by the spatial acts of displacement and deracination;[4] and as a new literary genre conditioned by the uncertainties of colonialism.[5] However, it has never been read from a Dalit ecological lens, in which different components of the earth appear as subject, self, person, life, death, and consciousness, in the everyday life of ex-untouchables. It is generally thought that Dalits, alienated from nature and deprived of access to natural resources, cannot be immersed in their environment. I have earlier raised the question of why there has been no recognition of ecological underpinnings in the writings of subordinate castes by the wider canon of environmental literary sphere. Yet, Mallabarman, in the midst of water passages, reflects on the magnificence of the river:

> Dawn comes on the river with the most exquisite beauty. The sun is not yet out and the transparent sky, taking on a hint of blue, spreads bluish creamy white light throughout. The sweet tones and soft serenity of this open expanse make all the pores of the heart sing with joy. The unobstructed gentle air rouses the delicate clapping of hundreds of millions of baby waves.[6]

Mallabarman journeys through river and water, banks and ghats, land and people, fishing and farming, rain, storm and flood, community, caste and conflict, and songs and festivals. Autobiographic in a wider sense, the Malo writer weaves a complex narrative of nature, space, place, time, and community at the turn of the century. Along with a human narrative, another memory lives in the novel, a passionate and unquiet one: it is the memory of the earth itself and what is inscribed upon her in time. Introducing the novel, Bardhan has remarked:

Vivid water, sky, and landscapes of seasons and human rituals mingle and converse happily or sadly. Images change and moods shift, but so deftly blended with the face nature presents, with the intricacies of festivity or crisis, and with the nuances of antecedent events that even the mundane and the magical, the gruesome and the enchanting all become natural.[7]

Titash is not about the nature of the natural world or of living things. However, it contributes immensely to the understanding of the natural world and people's lived experiences in environment. Through an eco-literary reading of the novel, this article underscores that in Dalit narratives, the earth can be perceived as a bearer of the 'worlded' worlds, as a reservoir of traits, and as a locus of individualization, which also gets scripted as community, gender, and caste consciousness. Drawing from the discipline of eco-criticism,[8] I bring out the profound caesura experienced by Dalits with nature. It contemplates what brings people together and what pulls them apart in their 'traction' towards nature, and how culture, gender, and caste acquire an affirmative role through their varied moves in nature. The political and economic history of Titash and the region in which the river is located are not dealt here.

The Journey of *Titash*

Even more than other regions, Bengal witnessed significant social, cultural, economic, and political changes in the early twentieth-century colonial context. The introduction of a new land tenure system, advent of early industrialization, emergence of new urban centres, rise of Bengali educated middle classes, and the incipient beginnings of nationalist movements were also intertwined with nature and environment, with implications for villages and rivers.[9] Alongside these tumultuous developments, there were also the incipient beginning of Dalit identities. Historians have highlighted the emergence of layered movements among 'low-caste' groups such as the Rajbansis and the Namasudras in northern and eastern districts of Bengal.[10] *Titash* frames itself amidst such developments.

The novel, divided into four parts, with two chapters in each, begins with images and narratives of the river Titash as a living being. The river, inscribed in the history of time and geography, is perceived as eternal and infinite. The Malo community lives with Titash's waters, waves, banks, boats, and fishes, and their joys and sorrows are blended with the changing course of the river. The beauty of a flowing river, the exciting water journey of a young fisherman Kishore, the labour and love of catching fishes in a boat, and Kishore's marriage are all suddenly shattered during his return travel, when robbers take away his wife. In deep pain and shock, Kishore becomes mad.

In the second part, a young woman – Kishore's wife, referred to as Ananta's mother – ventures out to search for her lost mate and father of her child. In her new life, she forges bonds with a new place and people, especially Basanti, and dips into folk cultures and festivals. However, the flow of life gets disrupted by the painful deaths of Kishore and Ananta's mother in strange situations. Ananta becomes an orphan.

The third part is about Ananta's quest and search for the meaning of his life through education and knowledge. Rainbow characters and episodes – anchored in water and land, fishing and farming, boating and labouring – have a river-like flow, where Ananta tries to carve out a space for himself, amidst a criss-cross of religion, caste, class, and culture.

The fourth part witnesses the emergence of a rebel woman, Basanti, and her spirited struggles to reclaim her dignity and pride even in the wake of severe adversities. *Titash* ends with changes and disharmony in the river, appearance of alluvial land and land conflicts, ruin of river and community, and ruthless march of an urban culture.

Moving along interlinked levels, the novel portrays the long life and culture of a river and the Malos. The river remains the lifeline of the community and their livelihood, until the time the river water is alive and running. At one level, along with river and land, youth and women forge their bonding, begin to reorganize their physical and social worlds, and try to carve a future for individuals and community. At yet another, social groups reflect a set of economic, social, cultural, and religious views. At a different level, within the lifetime of a few generations, the earth appears barren and thousands of bonds linking humans with nature disappear. It appears in the end that the world of Malo, having gone through the anguish of the world of nature, can never be the same again. In the following sections, I take four features of the novel to analyse the depth and breadth of Dalit life in nature. This brings us closer to an understanding of how Dalits live and perceive their lives on the earth, and their relationship with nature.

Earth and Life

Titash is a river. Rivers can be geological wonders and offspring of a natural past. In a complex, dynamic ecosystem, a river becomes central to the environment of a region. Its lineage is deeply intertwined with the life and times of humans and non-humans.[11] Rivers have been visible and major symbols of nature–human relationships, displaying a complex web of independence and dependence. However, Titash presents a deeper and wider cognizance of unity and continuity between the earth, nature, and environment. In Dalit imagination, a river encompasses the earth and flows without any rigidity of boundaries between different elements of nature. The ever-running river crosses over the subject–object, mind–nature, and human–nature dualism, with limitless creativity, agency, and autonomy. There is a discovery of the earth in river, river in human, nature as a site of culture and religion, and environment as an everyday experience.

I draw my theoretical framework here from the cultural historian Thomas Berry and African American environmental author Carl Anthony, who weave a new story about who human beings are in relationship to the story of the earth. In his book *The Dream of the Earth* (2016),[12] Berry notes that one of the remarkable achievements of the twentieth century is its ability to tell the story of the universe from empirical observation, with amazing insights into the sequence of transformations that have brought into being the earth, the living world, and the human community together. According to him, the earth consists of a 'communion of subjects not a collection of objects'. The earth and its inherent powers bring forth a marvellous display of beauty in an unending profusion, which gets overwhelmingly transmitted to human existence and consciousness. Visualizing the future of earth–human relations, he says that 'our own dreams of a more viable mode of being for ourselves and for the planet Earth can only be distant expressions of this primordial source of the universe itself in its fullest extent in space and in the long sequence of its transformations in time'.[13]

Reflecting on African American environmental history, and its relationship to the earth, Anthony states that the knowledge of the earth and the place of human beings in it gives a sense of identity and belonging to the black people. He narrates:

> The earth is the ground we walk on, the sea and air, the soil that nourishes us, the sphere of mortal life, the third planet in order from the sun, near the center of Milky Way galaxy. Everything that we do, or aim to do, is governed by our relationship with the earth – to its inspiration and resources, to our consciousness of its relationship to the cosmos, to our affinity with human and other-than-human life. Our knowledge and affinity with the earth, in all of its richness of life and diversity, stretches from the tiniest particles, waves and cells, to its plant forms and ecosystems, its rivers, mountains, and seas, to the majesty of our solar system, galaxies, and outer edges of the universe.[14]

Titash begins with brimming water, its surface 'alive with ripples', its 'heart exuberant', as she has been for hundreds and thousands of years. The origin of the river, the history of its evolution, even the etymological source of its name never come into question. The river is a beautiful evolution, figuratively and systematically. Its presence in space, its creation of boats, fishing, fishers, people, waves, winds, strokes, and storms, its relationship with weather, sun, moon, water, fish, and bank – all are living testimony to the river's abundant existence and freedom in her journey on earth. Even the feverishly hot mother earth remains cool under the water of Titash. However, the river's evolution is complicated. The links within the river system, different elements of nature and environment living within the riverine as inter-dependent units of the earth's biosphere, human and non-human – everything demonstrates the immense plurality of individual beings in nature, while also showing their intricate interconnectedness. In *Titash*, different narratives about the river recognize her complex diversity on earth:

> The course of the river Titash curves here like a bow.
>
> From one season to another the waterscape changes colors and shapes. Now, at the start of rainy season, its misty soft colors resemble a rainbow. Green villages line the two sides of the whitish water. Rain falls continuously from the soft grey sky above and rainwater mixes with farmland soil to run in hundreds of brown streams into the white flow of Titash. Together, they create an atmosphere of enchantment, of sweet spiritual rapture, like the world inside a rainbow.[15]

Titash is an open space. She is flowing and vibrating in an earthy environment. She is riding on the land of the earth, nurturing earthly food, making earthy beings survive on her water, absorbing the earth's atmospheric twists and turns. Her body functions in tandem with others on earth. The human and non-human take a piece of the riverscape and support their life through it:

> Titash is so gentle. Even on nights of rain and wind, with their men out in Titash, the women do not really feel afraid. The wives can sleep, imagining their husbands nestled in their arms. The mothers can rest, imagining their sons in their boats on the gently rocking Titash, peacefully gathering up the nets filled with fish.[16]

In *Titash*, every subject is equal and important. All organic and inorganic substances long for their natural life. Different substances manifest themselves at different places but proceed and culminate according to their natural origins of river, water, life, flow, and freedom. All along the course of the river are ghats that 'slope gently to the water' and are alive with pictures of life. The banks appear vivid, 'imprinted with stories of a mother's affection, a brother's love, the caring of a wife, a sister, and a daughter'. The water of the river is 'unruffled and transparent'. The constant sound – *jha–jhajhim–jhim* – of rain comes on the river's surface. The wind is a 'wanderer' like water. The rainbow in the sky is immense: 'Spanning two far corners of the sky, it arches high above with its seven colours, each so distinctly defined from the next with no blurring!'[17] The moving fishing boat is like 'a water snake heading home at dusk'. The fish dances freely and the beauty of the first fish catch is as fresh looking as the dawn of early spring. The village is shaded by the growth of green plants and trees. 'True' are the people who live along the riverbanks. Villagers have not only nets but also ploughs. The presiding deity of their life is 'Adamsurat, the half-made, half-female guardian of the sky, pointing two fingers in different directions, one to the brimming river and the other to the smiling fields'.[18]

However, such animate subjects also start moving far and further away from the riverscape, of whose components they are made of. Eventually, they cease to exist naturally. The moment water, people, and community leave their river-found nature, they enter a new stage of their existence. For this reason, who can tell what might happen to the river itself, from which everything on earth is so vibrantly created, after prolonged exposure to non-river and non-earth conditions. What new qualities she might acquire and what qualities known to her might get lost. It is a seized river now. On the whole, river and river lives, despite their efforts to adapt to alien environments, find themselves in a paradoxical situation. Even when the river gives away her space to land, agriculture, and conflicts, there is no escape from death:

> Stuck out on dry ground under the sun, the fishing boats crack; too little water in the river to keep them afloat. The Malos still do not give up fishing. With the triangular push net on one shoulder and a narrow-neck fish basket on the other, they desperately roam all day from one village to another looking for a pond or a tank. When they spot a clogged pond in some village, they scan it with hawk eyes. Their bodies are skin and bones, their eyes sunken in their sockets.[19]

In *Titash*, the river is undoubtedly the most complex organism existing on earth. In different spaces, seasons, and journeys, the river continues to live with her native biosphere – the environment and people of her existence – and this environment also fences her off from destruction. The river and her elements have something in common, like characteristics of freedom and choice, subjectivity and creativity, which also create the Dalit environmental persona. The river and human beings drift apart from the earth's natural environment, and its biosphere, which has given birth to them and reared them. The earth nevertheless always remains with the Malos, accompanying them on the uncertain routes they have chosen. However, there is no river left by then: 'Titash seems like an enemy; turned hostile and merciless. It has become a total stranger today.'[20] I will now deal with Malos, the river people.

Malo and Maloness

Titash is about Malo men and women who live besides the river. Malos are organically integrated with the body of the river, her life, and health. Their integration manifests in direct and indirect forms – through fishing and farming, and food, water, festivals, songs, and recreation on the riverside. Malo is also a settled and territorially organized society, a Dalit community with everyday concerns of environment, labour, and livelihood, and a cultural community with a unique blend of Vaishnava *bhakti* (devotional sentiment) and Sufi spiritualism (mystical Islamic belief). It is a complex and collective formation of village, homes, occupational and social groups, animals, plants, floods, migration, and mobility. However, in *Titash*, the Malos are first and foremost human beings. They are also not isolated or separated from other subjects. They are always already in the river world, inside spaces and times with which they have identification. I explain the being of Malo, drawing on some of the formulations put forth by the influential philosopher Martin Heidegger in his *Being and Time* (1966).[21]

According to Heidegger, being is time. The being of a human being, who lives finitely through a stretch of life and death, is defined by unity of a three-dimensional time: a past, moving through a present, and available for a future of possibilities. Being, as 'an existential analytic', is faced with the crucial question of 'to be or not to be?' This questioning and thinking about 'who' has been shaped by a personal and cultural history of being. 'Mineness' – as a matter of oneness, of one's own experiences, of authenticity that acquires a certain common structure in one's average everyday existence – is an integral part of being. Being is constituted in a world of things that are useful, meaningful, and humanly significant and valuable. Being is fully immersed in an environment of 'being-with' – a common world of things, persons, and social practices, which are experienced together with others:

... Da-sein understands itself-and that means also its being-in-the-world-ontologically in terms of those beings and their being which it itself is not, but which it encounters 'within' its world.[22]

Heidegger also analyses the human being through concepts of 'state of mind', 'mood', and 'throwness'. The human being in its being-in-the-world is always caught in the throw of various moods – feelings of calm, peace, fear, and anxiety. Anxiety is the ultimate state of being which is of 'nothing and nowhere'. However, a human being, through experience and freedom, has the potential to throw off the thrown condition in a concrete situation. Being has moments of visions and actions. There is a call of conscience that silences the chatter of the world and brings one back to oneself:

Da-sein has a mode of being in which it is brought before itself and it is disclosed to itself in its thrownness. But thrownness is the mode of being of a being which always is itself its possibilities in such a way that it understands itself in them and from them (projects itself upon them)....The average everydayness of Da-sein can thus be determined as entangled-disclosed, thrown-projecting being-in-the-world which is concerned with its own most potentiality in its being together with the 'world' and in being-with with the other.[23]

In *Titash*, Malo is present and active everywhere. Yet, Malo is not the one who lives in brick houses with walls around or travels on paths that reach towns and larger villages. Their only road lies in the river's midstream and boats are their companion. The river overlooks their activity and her presence constantly mingles with their daily chores. However, Malos also have a life of their own, individual meanings of their breathtaking being, which acquire significance from the world of things, from the actions and reactions they provoke, and from the manner in which they give expression to cultural and social values. This is a distinct, authentic oneself – Maloness – that comes out from their everydayness.

Maloness is the 'way to be' in the present, reflected through experiences and actions. It comes out intensely in the 'the stories of a mother's affection, a brother's love, the caring of a wife, a sister, and a daughter'; in the bodies of Malo men who have 'the strength of elephants'; in the 'vivid pictures of genuine people of flesh and blood'; in 'accounts of their humanity and their inhumanity'; in 'long fishing trips' beckoned by the mysteries of invisibles and dangers; in 'the memories of the flower woman'; in occasions of birth, death, and marriage; in people becoming the wish-tree; in floaters, butterfly, and rainbow; and in Kali and Mansa (goddesses) worship. The river is the closest and most obvious element of Maloness. Ananta, in the river, is thus portrayed:

> All his sensations are absorbed in the river that flows around him. Ceaselessly and as lightly as gossamer, the river draws out all his feelings, all that is the essence of his existence. Forgetting both past and future, he floats in the continuous flow of this freshly awakened, eloquent present. It is as if the real journey of his life starts here.[24]

On earth, the Malos discover themselves in different states of mind, moods, and thrownness. The diverse elements of nature provide the foundational ways in which the Malos can be elegantly located. They have, firstly, a there-ness in their world. They are attuned to the natural and physical environments via the vibrations of their bodies, tools and fruits of their labour, and souls of their passion. In there-ness, young Kishore and Subal can float their boat, lower their net into the water, perform the ritual of starting their fishing day by taking a cupped palmful of river water in their mouth, and enjoy the sight of their first catch of small slim silvery fish. In the boat's swaying motion, 'Ananta could step from the boat onto those paths. But those paths lie in water; only fish can travel there.... Ananta's mind becomes a fish and dives into the water.'[25]

The Malos' world is also thrown into different moods of fear, uncertainty, pain, danger, risk, tragedy, and death. Completely immersed in river, water, and land, their state of existence and moods become the essential markers for understanding the Malos' existence on earth. At times, young Kishore consoles himself and also encourages his fellow fishermen not to be afraid in the river water after hearing the faltering rhythm of the scull: 'Don't be afraid, Subla. Danger's always near the banks, nothing to fear in midstream. Mother Ganga (river) reigns there and protects boatmen from harm.'[26] Ananta lives with the anxiety of whether the river's current would take him to distant lands, the waves would rock him, and there would be no one else around in the surrounding darkness. There is fear and anxiety of heavy rains, storms, and floods, which often bring great loss to the Malos. There is deep sorrow and pain about getting separated from water, which makes the Malos gasp like a fish.

A Malo being is bound by finite time, often ending in death. In *Titash*, death is a subject of frequent occurrence among the Malos, which culminates in the death of the river. In the first part of the novel itself, Kishore is shocked at seeing the floating dead body of a female. The Malos, in their different states of mind and mood, realize their finitude in death: 'Those who are born must die some day. So each household is bound to be visited by death.'[27] Most of the deaths are not anticipated and come suddenly. They are relational and ignite silent reflections about one's life in family, society, and nature. Death causes individual and collective pain; it is also a call of conscience. Sibal dies in the river when he jumps in the flooded water to save the boat. Sibal's wife 'no longer grieves for her husband; the pain she felt has subsided. The thought of the terrible death her husband died comes back to her from time to time. She tries to imagine the scene.'[28] Death does not liberate the Malo, but it shapes their self. A certain Maloness sees the death coming to them through the death of the river, as close as a son, husband, or friend. The relation to death is not only about one's own; it is equally about the grief and demise of one's nature being: 'Those who died are in a way saved. Those who are alive only wonder, how much longer! From the side of Titash, the answer seems to drift in, not much longer!'[29]

The Malo lives in a time where the river on earth has a multiple, non-dualist existence. The temporality of a human being is not confined to the present but flows from a past into a future. A distinct Maloness comes out from a distinct river life, and both share some common values – free, flow, uncertainty, living, and dying. However, taking a leaf out of Heraclitus, no Malo ever steps in the same river twice, for it is not the same river and he is not the same man. Movement and change are ingrained in Dalit environmental imagination. I will now deal with the river spaces and places that characterize Malo men and women.

Spaces, Places, and Commons

Titash has several stories of spaces, places, place imaginations, and practices of the Malo community. In the beginning, the river appears on earth as an unlimited, free space. The river space itself is a key marker of earth and time. The Malos create places out of river spaces and give them their distinct look, feel, and ambience. These places – villages, hutments, banks – in turn are inseparable from the spaces they have evolved. The Malo community as a whole have their space–place attachments and associated practices. Maloness is based on a unity between space, place, and society. In the course of their river journeys at different times, Malos' activities and practices also present Dalit imaginations of rivers, commons, and community. They also challenge the under cables of environmental and social controls that are laid to work in the river areas through several internal and external mechanisms. At the same time, Malos have a tremendous ability to move into new places and make them of their own. To reflect on spaces, places, and Malo commons, Buell's works on eco-criticism, and J. T. Roane's[30] scholarship on black communities' relation to geography, sexuality, and religion are important.

Discussing the imagination of space and place in environmental criticism, Buell talks about a five-dimensional phenomenology of subjective place-attachment that together makes possible a 'critical grasp of place as subjective horizon'. Place consciousness and bonding evolves from an orientation in space, as well as a temporal orientation. At the spatial level, there is a different mental mapping: traditionally, it has been a strong emotional identification with the home base from which most of one's life is led. However, this changes under modernization, when homes can be spread at different places. Imagining a place can become an important element in place attachment. It also has temporal dimensions. Conversely, place itself changes. It 'is not entitative – as a foundation has to be – but eventmental, something in process'.[31]

Roane examines black communities of the lower Chesapeake Bay, a unique meeting of water and land, river and farm, in Tidewater Maryland and Virginia, in the antebellum and post-emancipation periods, and terms their engagements and practices of natural resources as the process of constituting 'Black commons'. He refers to such practices of place and alternative figuration of land and water as 'plot/plotting': 'the various iterations of a cosmological, geographic, and social outlook with material and political manifestations', through which a community claims and creates a set of communal resources.[32]

Plot is the centre of a number of distinct and overlapping activities of Blacks, which foster a vision of de-commodified waters, landscapes, and resources. Foremost, the plot becomes the site of an elaborate funeral as well as burial of the loved ones. Plots for the dead signal a key method of organizing community, 'anchoring the present through the past within the grooves of a landscape'.[33] The plot signifies the garden parcel – spaces for land usage organized through use value, and sustainable biological and social existence. The plot also leads to the extension of such use value into the region's forests and waterscapes. Roane concludes:

> The plot signifies insurgent cartography whereby enslaved and freed people used the other modes of plotting to articulate geographic identities laden with epistemological possibilities and horizons for the future outside the parameters of white dominance and control and through the ecstatic, beyond the theology of dominion.[34]

In *Titash*, Malo fishermen are always on the move and, during their long journeys on boats, often anchor themselves at several places, and stay over time with people from their community. However, on the riverbanks, the Malo community as a whole, and women especially, have heart-thumbing everyday lives. Malo men and women inhabit villages, where they live, work, play, and worship. They migrate to new Malo places. They imagine new places and even when they are slipping away from the river zone in changing situations, they have a sense of 'no place'.

In the boundless river, Kishore, Tilak, and Subal travel for days and halt in a settlement of Malo households where they discover different curves of river and riverbanks, new varieties of larger fish and distinct tastes of cooked fish food, singing of Vaishnava invocations and celebrations of Dol Purnima (festival dedicated to Hindu God Krishna), and full moon of the swing: 'singing, playing with color, and feasting, from morning till night'.[35] In one of these places, Kishore finds her bride: 'your partner in work and in virtue, your mate in this life and in the life after'.[36] After getting free from pirates' captivity, Kishore's wife (named Ananta's mother) finds herself in new places, where she develops close ties with fisherwomen. She works hard at home and outside, filled with solidarity and hope. She enjoys the happiness and colour of feasts and festivals, 'for the realisation of good possibilities'. Ananta finds a new place some day and experiences his world transformed:

> Under the star-filled clear sky the river lies motionless. In the distant vault of the sky, stars in innumerable clusters form luminous pathways. How immense must be the beauty of those paths, and the joy of walking along them, treading on the star-flowers, and watching the countless other star-flowers on both sides and overhead![37]

Titash ends with the loss of place – the river is shallow, the Malo families have left the village, the Malo neighbourhood is no more, and the empty huts are covered with wild growth. They have been pushed away by their situation: 'No space for you, no place for you!'[38] In Malos' everyday life, place-attachment-and-movement creates a scenario where they are in a flux and are never fixed. The place attachment is not a uniform, linear, and infinite series of spatiality.

For Malos, the river Titash, its tributaries, and canals carry multi-layered and complex social, economic, cultural, and natural values. They embody a dynamic interplay between social integration, cultural diversity, and economic usage, and are crucial for social, spatial, and economic transformation of society. Titash is a fleeting common in a fleeting social world. Every Malo home has fishing boats and nets to harness the river water. Flowing and living by the river, it is their centre of life and livelihood:

> The clear water is transparent in the sunlight and the current is gentle. Bands of *katari* fish float up to the surface and play. Other small fish gather under the film of oil from the bodies of father and son; they prick the film with tiny bubbles. The fish come so close that Romu extends his hand to catch one but can't – they are always too quick.[39]

Titash, as Malos' common, is also a space for myriad and vibrant community activities. It is the place of a boat race, when edges of water teem with covered boats, and each is packed with people. The race boats roam all over the wide, open stretch of the river, their oars swinging unhurriedly with many different melodies of songs. The riverbanks are also still open spaces for the anchoring of boats, and for the drying and weaving of nets. Sometimes, the Malos have a bath here, and worship the sun with water, wildflowers, and tender blades of grass, chanting: 'Please, Sun God, take this sacred water I offer; Seven cupped handfuls I carefully measure.'[40] Beyond the active banks and ghats are river lands, where common crops, trees, and shrubs are stretched far into some distance. Malo fishermen and farmers go about together ploughing the water and the land. Kadir and Banamail, peasant and fisherman, think that 'peasants and fishermen have a link nobody can cut even by hacking or erase even by scrubbing'.[41] These common activities in the vicinity of the river portray elements of the natural environment, but their meaning stretches in Malos' vision to signify the essence of human society and what it ought to be.

The river common creates a ground to build a lasting human common. Malo people – cutting across age, gender, and religion – forge a deep bonding around their river lives. Young Kishore and Subal, Ananta's mother and Basanti, Ananta and Banamali, Basanti and Udaitara live with immense and courageous human solidarities in their common natural and social life. As the translator states: 'In *Titash*, the scenes between old Kadir Mian and young Banamali, between Baharullah and Ramprasad, between Romu and Ananta, between Jamila and Udaitara are some of the most moving and humane portrayals of affection that transcends religious barriers.'[42]

Malos, in river spaces and places, show a tremendous human capacity to live their lives in both local and migratory ways. This in a way is a reaffirmation of the river life – moving freely to different places. The life of the Malos – work, landscape, and values – is also realized in the making of commons, both as a place of resource use and as a life of resourcefulness and solidarity. This is the case, as Roane argues, with black commons – an extension of new modes of human connection, belonging, and reciprocity, and a different version of earth stewardship. I will now focus on culture, gender, and caste, which plays a significant role in creating Malo commons and Maloness.

Culture, Gender, Caste

In the Malo neighbourhood of Gokanghat, young Mohan contemplates his cultural life, and how his songs, stories, and sayings are woven into Malo festivities, religious celebrations, jokes, and riddles, which have a distinctive beauty. Songs expressing sweet feelings in beautiful melodies are passed from generation to generation. However, he feels that 'people other than Malos themselves have no easy access to the heart of that culture to partake of its nectar, because the Malos' way of lyrical appreciation is unlike that of all the other communities. What the Malos have woven into their hearts, others treat with ridicule'.[43] Along with the river of life, livelihood, place, imagination, and social bonding, the Malos have a dynamic cultural life: they have their own socio-cultural networks, they organize community events; they love songs, dances, festivals, ceremonies within their community; and have a distinct religious identity. In *Titash*, the Malos' cultural actions appear regularly, not always in a fixed frame of traditional annual Hindu festivals, but in a process of river folk life, with ever-changing weathers and months. The spectrum is created through an interaction of human, cultural, and natural processes. Freedom in river space strengthens the sense of an autonomous cultural consciousness, as they are participating in a variety of natural and social worlds, without inhibitions and self-containment.

Wonderful, different *baul* songs (folk songs composed by both Hindus and Muslims, and sects of wandering singers known as the *baul*s) originate every day from the Malo fishermen moving on boats or riversides. There are *shari* (group songs during the harvesting of paddy and the rowing of boats), rowing, love, work, and social songs, evoking different moods of life. Their themes are not confined to a specific place and time. They have divine, eternal, and spiritual elements within the references of Malos' living world:

> In life's immense waters at cosmic play is the pure.
> Rocks float away and rafts capsize on dry shore![44]

A lone boat and the oarsmen rowed and sang in a chorus their *shari* song:

They all have their own ones, but I have none,
Thundering inside me are the waves of the ocean
I came to the river's edge hoping to go across:
An empty boat with no boatman helplessly bobs.[45]

The Malos live in many villages and have different beliefs, customs, and institutions. However, such differences are in dialogue with each other. In spite of the spread-out physical locations of their villages, their cultural practices share some common virtues and ideals, and manners of self-expression and collectivity, which have a distinct pattern. In Shukdebpur village, people gather in a singing session after the evening has settled in, comprising both Krishna and Shiva (Hindu gods) followers. After singing a Vaishnava invocation, Tilak sings a song about the ascetic Shiva, dancing with his horn and taboret in the cremation ground in the company of spirits. This song does not quite appeal to the Krishna devotees. Upon their turn, they sing about the eternal union of Radha and Krishna. They have puffs of ganja (hemp) and sweets in between, and all disperse in sometime. Such singing sessions occur in different villages at different times:

O, what a beautiful sight it's now becoming!
Cowherd boys of Brindaban restlessly calling.[46]

The annual Kali worship and Holi festival have a natural and physical touch of the Malos. For a grand Kali worship, a boatload of clay lands at the ghats besides Titash. Malo boys have to work to soften the clods of earth to pliable clay, by stomping and treading on it, dancing and walking in it. The craftsmen, with the support of Malo women, have to spend days applying the smooth clay layer upon layer to construct the body. After days of work, when the image is ready for worship, women have to fetch water from Titash for the worship. They also pick and arrange flowers, prepare offerings of food, and place them before the image. On the day of Holi, Malo fishermen do minimal fishing, and retrieve their nets early in the morning because it is a day of singing, to greet the arrival of spring, and for the renewal of river. Someone starts with the beat of tom-tom, 'O Spring, you've come with joy, but my *lal*, my red one, has not come!'[47]

In the cultural and natural sphere of the Malos, women have an overarching presence. An integral part of Malo society, they appear independently, as well as with men. They exist as human beings with their distinct spheres of physicality, appearance, labour, love, and loss. *Titash* does not portray them as 'mother earth', inheriting some essential qualities in relation to nature. Their experiences and practices have a complex relationship with the everyday life of Malo society, where they have their own space, time, and agency. In the context of rural African American women, environmental historian Dianne Glave shows that they developed their relationship with nature by way of gardens they had grown as slaves and then as freed women. They developed their expertise by drawing from community knowledge and their own interpretation of agricultural methods. Women and men often supported one another through complementary roles and strategies designed to support the family unit. She explains:

> To plant their flower and vegetable gardens, African American women used their hands – darkly creviced or smoothly freckled; their arm – some wiry, other muscled; and their shoulders and backs – one broad and another thin. They dropped small seeds into the soil with their veined hands. They wrapped their arms around freshly cut flowers to decorate tables in their homes. They bent their shoulders and backs to compost hay, manure, and field stubble, and transplanted plants from the woods into their own yards. These women developed a unique set of perspectives on the environment by way of the gardens they grew as slaves and then as freed women.[48]

Malo women of various ages are thoroughly grounded in multiple activities in the entire river area. They are found in birth, marriage, and death, in river and boat, in land and agriculture, in festival and ceremony. In Shukdebpur village, after the songs of Krishna and Shiva followers, women form a dancing circle of Dolmandal, their hands poised, with some playing finger cymbals, and some clapping in rhythm: 'The cymbals strike out all at once, filling the air with the beat *jha-jhajham-jham* and the sound of innumerable bangled hands clapping. The women set their feet to dancing together.'[49] Like men, Malo women share their life stories amongst themselves. Mother, daughter, mother-in-law, daughter-in-law, sister-in-law – all narrate their intimate home stories by the ghats or riverbanks. Outside home, they often have collective working practices. Spinning and reeling all available threads, fine, medium, or thick, for fishermen's nets; cleaning up fishes; identifying and organizing them for the market; going to the market; grinding rice from the new harvest – several such activities are 'Malo' and 'fisherfolk' in nature and have a critical value in the entire chain of fish cultivation and use. At the height of fishing season in the spring, Malo women and men from many different villages gather at Shukdebpur village, for weeks and months together. This is the time when thousands of large fish are caught. Men bring basketfuls of freshly hauled fish and upload them before the women, and women's hands work with the smooth swiftness of machines, gutting and cleaning large fish. This is the process for drying fishes, to sell off to fish traders. These everyday interactions of Malo women with nature are not something 'out there'; they are 'in here' – close to their heart and body.

Nature has its darker side too. Malo emotions are severely shattered under water. Malo women know how treacherous Titash can be in times of floods, storms, and heavy rains. They know that this will come one day and will entail risks and vulnerabilities. Women's individual and collective lives bring forth time and again the chaos and disorder in nature. The capriciousness of water, waves, and winds are unsettling for them, their homes and families, and their community. Nature is indifferent to human sufferings; humans are not. The rainy season brings the hardest time for Ananta's mother. They have nothing to eat. However, Subla's household is always there for her. An old Malo woman faces the storm's fury in full force – roofs of huts are damaged, trees are uprooted, boats on riverbanks are smashed. Each moment is frightening. Still, the old woman, concerned about the village, starts shouting at the storm. However, the storm is unquestionably powerful and unperturbed. Pitching her voice at the highest level, she shouts again at the storm. Ananta is amazed at the power in the woman's voice – so much force in her command:

> Go away, fellow, go to the hills and the mountain tops!
> Go and fight on your way out with the big tree tops![50]

Malo's cultural world is also interwoven with the Hindu caste system, where they are placed at the bottom of the four *varna*s. Malos are Shudras. I have argued elsewhere that Dalits' rich cultural corpus reflects the complex interconnectedness of caste and nature. Dalit experiences and narratives constantly underline their everyday ecological burdens in a marked hierarchal order: living with nature, they are constantly negotiating with, and challenging, caste domination, while simultaneously articulating their environmental imagination. Yet, the Malos living around the river Titash, mostly within their own community, actually possess an autonomous cultural domain, which is vibrant enough to mitigate caste exclusion in their everyday existence. The communitarian river life of Dalit women and men, their cultural memories, and religious practices have historically generated the necessary ingredients to create counter-narratives to caste domination. The nature-driven qualities – freedom of Titash, equality of different subjects, Malo commons, openness of space and place – have provided the bedrock for a 'casteless' culture in their vicinity, even when caste is not eliminated.

Caste is, of course, present in Malos' memories and they are surrounded by caste realities. However, it is not at the centre of their existence, and their identity is not defined by the Brahmanical social order. Malo women refer to the poverty-stricken state of Shudra people. In their boat journeys, Malo men construct images of Brahmin and Kayastha women. They know that communities in other areas do not let Malos enter their homes. They consider them polluted and hold them in contempt. They face Brahmin fish merchants and landlords, who try to capture their produce cheaply. Still, caste does not centrally determine the lives of the Malos around Titash.

At the same time, Malos' caste–cultural narratives are prominent and decisive. They narrate a long story of a Brahmin, Buruj, who goes on a long journey. In the parching heat of the sun, and after growing tired from the strain of walking, Buruj is looking for water, but he finds no river or pond anywhere. He finds a nice hut, with marks of sandalwood pasted on the door. He thinks that such a clean and tidy home cannot be of anyone other than a Brahmin. He drinks water there and, refreshed, asks the maiden her caste. She replies that they are of the peasant-gardener caste. This also means that the Brahmin Buruj has now lost his caste. He loses his caste in a gardener's home. Losing his Brahmin-hood, Buruj now lives in the gardener's home as a gardener.[51] In another plot, when Ananta's mother is asked her caste in a new place – if she is the daughter of a Brahmin or a Kayastha family – her answer is that 'she was a fisherman's daughter'.[52]

The river is not a segregated space. It is marked by culturally affirming activities of Malo men and women. In closely and critically accessing nature, Dalits produce social and cultural correlation and causation. Such an interrelationship, based on everyday cultural affirmation, has often found expression in colourful celebration and creative communication. Cultural affirmations are not gender and caste exclusive. Rather, the assertive cultural and gender identities of Malos blunt caste and religious boundaries. Malos came out cognitively and socially resilient from such an affirming environment. Malos' cultural treasures, from *Dol Purnima* to *baul* songs and from Holi to worship of Manasa, bring communities together across lines of difference and ignite Malos to achieve higher levels.

Since the mid-twentieth century, siltation of the river and the formation and spread of silt-beds has intensified the transformation of Titash. Vast growth of land coming out of silt-beds, emerging issue of land ownership, and development of agriculture on the one hand, and the opening of river and water for different users, including the emergence of fishermen's associations and fisheries administration on the other, have severely impacted the river areas. Titash is not a 'divine' river. Significant natural changes in the river's character, and its bearing on people, have been seen as the decaying and dying of the river system and the lives and cultures woven around it.

Malo men and women were ordinary fisher people, who always lived in villages and cultivated strong community ties. Individually and collectively, they were doing open, fresh water, small-scale, and migratory fishing full time. Only a few of them were engaged in part-time farming. The Malos' labour was enormous, skilful, mindful, and risky. The identity of the Malos was based on their everyday interaction with the river, water, boat, and fish. The intermediation between these elements, since one is accomplished only by means of the other, was never ideal. However, they were constantly on the move, to an infinite horizon – an upstream, new tributaries, and long distances. River fish was their livelihood and food, enough to satisfy existing wants. The changing weather, water, rain, and flood was always bringing surprises, as well as new fishes into being. Every stage of Malos' life – birth, marriage, death – developed alongside, and in interaction with, nature. However, in the times of the drying river, Malos, their neighbourhood, and community seem to be dissipating.

The natural and human development of the region confirms a pattern of characteristically degraded landscape, defined by development, inequality, conflict, and hopelessness, imposed over calm ecologies. Yet, the Malos, even as vulnerable inhabitants of the river, provide dialectic visions of Dalit environments. Dalits are craving for life on earth. In *Titash*, the river symbolizes it, and becomes the locus for ecological–social interactions. In the histories of life on earth, Dalits embrace the interactions of different elements of nature; the evolvement and habitability of ecosystems over the years; the free existence of water, lake, river, and land, and their freedom of movement across their surfaces; the wide variety of natural resources and species for survival; and the activities and changes in nature. The earth is a free space for Dalits. However, their 'earthy' existence also makes them realize that earth's biosphere is being significantly altered, that atmosphere has no definite boundaries and that their future is tied with multiple terrains, which vary greatly from place to place.

The idea and practice of a local commons is most endurable for Dalit environments. Commons are not a singular entity; they are complemented by plural characters. Titash was an open common, coexisting with riverbanks and ghats as also commons. Together, they signify resource sharing and use, under an open-access property regime. Such regimes are also subjected to production and exchange values, incomes, and losses, but they occur primarily at local scales and are operated by interactive community practices. Commons are accessible to everyone. They are used by all, owned by none. This gives Dalits a place, security, and confidence on earth. Dalits cherish, celebrate, and deepen the sense of commons – with nature commons and human commons often bonding together.

In Dalit environments, culture, women, and caste generate a rich repertoire of songs and music, gods and goddesses, myths and stories. They comprise a unique and different set of analytic, and the value system of Dalit environments can only be understood through them. They play an affirmative everyday role, and signify a memory, a metaphor, and a movement. For the Malos, Dalit environment often appears in a familiar terrain, underlining space and place attachment. And yet, they also have 'place-attachment-and-movement tracts', where they move dynamically to newer places. Freedom on earth, nature, space, and place keeps them moving. As they say: 'The past we can smile about, cry over, arrange our dreams around, but never get back. The present we want to hold in our hands. The future makes us string our hearts with hope. What is there to gain by dwelling on what happened in some long departed past!'[53]

In the following chapter, I will explore another significant counter-hegemonic narrative in the realm of Anthropocene. Specifically, I will examine the art, architecture, gardens, and sculptures created by Nek Chand, a low-caste Mali, within the modern city of Chandigarh. Nek Chand's creations are unique representations of the local geographies and the timescale marked by caste-based discrimination and industrial violence during a period of rapid acceleration in India. Through the lens of displacement, creative knowledge, and historic experiences; urban and industrial waste; and an acknowledgment of the diverse range of species and lives on earth and beyond, his artistic endeavours contributed to the discourse surrounding future ecologies and the recalibration of societal perspectives.

Notes

1. Adwaita Mallabarman, *A River Called Titash*, trans. K. Bardhan (Berkeley: University of California Press, 1993), 21. The Bengali novel *Titash Ekti Nadir Naam* by the Dalit writer Adwaita Mallabarman was first published in Calcutta in 1956. It was translated and published in English in 1993, with an introduction, afterword, and notes by Kalpana Bardhan. I have used the English translation.

2. Bardhan, introduction to *A River Called Titash*.

3. Adrian Martin, 'A River Called Titas: River of No Return', *The Criterion Collection*, 12 December 2013, https://www.criterion.com/current/posts/2990-a-river-called-titas-river-of-no-return, accessed on 20 January 2024.

4. Drishadwati Bargi, 'Understanding "Dalit Chetna" in Adwaita Mallabarman's *Titash Ekti Nadir Naam*, A River Called Titash', *Contemporary Voice of Dalit* 8, no. 1 (2016): 90–104.

5. Sravani Biswas, 'Nature and Humans in the Imagination of Bengali Intellectuals of 1930s–50s', *Studies on Asia* 1, no. 2 (2011): 15–34.

6. Mallabarman, *A River Called Titash*, 45.

7. Bardhan, introduction to *A River Called Titash*, 3.

8. Lawrence Buell, *The Future of Environmental Criticism: Environmental Crisis and Literary Imagination* (New Jersey: Blackwell, 2005).

9. Mario Prayer, 'Freedom in the River: Bengali *Bhadralok* Consciousness in Manik Bandopadhyay's *Padmanadir Majhi*', in *The Human Person and Nature in Classical and Modern India*, ed. Fabrizio Serra (Rome: Sapienza, 2015), 165–80.

10. Sekhar Bandyopadhyay, 'Partition and the Ruptures in Dalit Identity Politics in Bengal', *Asian Studies Review* 33, no. 4 (2009): 455–67.

11. Arupjyoti Saikia, *The Unquiet River: A Biography of the Brahmaputra* (Delhi: Oxford University Press, 2019).

12. Thomas Berry, *The Dream of the Earth* (Oakland: Sierra Club Books, 2016).

13. Ibid., 42.

14. Carl Anthony, 'Reflections on the Purposes and Meanings of African American Environmental History', in *'To Love the Wind and the Rain': African Americans and Environmental History*, ed. Dianne D. Glave and Mark Stoll (Pittsburgh: University of Pittsburgh Press, 2006), 202–03.

15. Mallabarman, *A River Called Titash*, 137.

16. Ibid., 16.

17. Ibid., 145.

18. Ibid., 44–45.

19. Ibid., 253.

20. Ibid., 245.

21. Martin Heidegger, *Being and Time* (New York: State University of New York Press, 1966).

22. Ibid., 55.

23. Ibid., 169–70.

24. Mallabarman, *A River Called Titash*, 73.

25. Ibid., 168.

26. Ibid., 35.
27. Ibid., 101.
28. Ibid., 120.
29. Ibid., 254.
30. J. T. Roane, 'Plotting the Black Commons', *Souls* 20, no. 3 (2018): 239–66.
31. Buell, *The Future of Environmental Criticism*, 73.
32. Roane, 'Plotting the Black Commons', 242.
33. Ibid., 242.
34. Ibid., 243.
35. Mallabarman, *A River Called Titash*, 48.
36. Ibid., 58.
37. Ibid., 168.
38. Ibid., 245.
39. Ibid., 193.
40. Ibid., 27.
41. Ibid., 141.
42. Ibid., 26.
43. Ibid., 233–34.
44. Ibid., 62.
45. Ibid., 219.
46. Ibid., 178.
47. Ibid., 130.
48. Dianne D. Glave, 'Rural African American Women, Gardening, and Progressive Reform in the South', in *'To Love the Wind and the Rain'*, ed. Glave and Stoll, 37.
49. Mallabarman, *A River Called Titash*, 54.
50. Ibid., 156.
51. Ibid., 209–10.
52. Ibid., 128.
53. Ibid., 101.

5

Anthropocene

'Modern' Chandigarh, Nek Chand, Rock Garden, and Dalitbahujan Anthropocenes

The term 'Dalitbahujan Anthropocenes' refers to the untouchable and low-caste population in the country whose bodies, labour, lives, and work have been shaped through their close proximity to industrial development and modernity for several centuries. This proximity is a consequence of historical colonialism and capitalism, systems of hierarchy and discrimination, polluted local environments, and persistent environmental casteism. It is rooted in the belief that Dalit and low-caste bodies possess a 'natural' capacity to endure and engage with the dirty and hazardous processes and substances of industrialization, with their labour being seen as essential to the nation's progress. In contrast to the dominant temporality of the Anthropocene, which focuses on a unified human society, the concept of Dalitbahujan Anthropocene disrupts teleological time and challenges singular narratives. It does so by recounting the experiences of Dalit and low-caste individuals within the caste–colonial–capitalist systems, expanding across various contexts such as agricultural fields, coal mining, bonded labour, displacement, migration, famines, and droughts – all in the name of civilization and modernization. Beyond the scope of caste and capitalism, Dalitbahujan Anthropocenes offer a creative alternative perspective on addressing issues related to extraction, waste, urbanism, and development. Nek Chand and his Rock Garden serve as counter-narratives, challenging the caste and Dalit blind spots within the Anthropocene discourse.

Nek Chand Saini (December 1924–June 2015) is widely known for his creation of the Rock Garden in Chandigarh city. Born in Berian Kalan village, 90 kilometres to the north of Lahore, and now in Pakistan, Nek Chand migrated with his family to the Indian side of the Punjab region after Partition in 1947. Until the age of 23, he, his parents, four brothers, and two sisters lived in an agricultural village of Pakistan. As a Partition refugee,

Nek Chand moved in several Indian cities – Jammu, Gurdaspur, Karnal, Panipat, and Faridabad, and finally settled down in Chandigarh.

Nek Chand Saini belonged to the Mali caste, which has been characterized in various accounts, and in various regions of India as an occupational 'Shudra' caste, traditionally and intimately connected with the occupation of gardening and cultivation-selling of flowers, fruits, and vegetables. The great anti-caste radical reformer Jyotirao Phule belonged to the Mali caste. In her masterful study on Phule, Rosalind O'Hanlon thus states:

> The Mali caste, to which Phule's family belonged, were gardeners. They made their living in western Maharashtra cultivating fruit, flowers, and vegetables, often using a network of kin connections to transport their produce to market and sell it. With this simultaneous concern with semi-rural cultivation and with large urban markets, the Mali caste often acted as a medium of contact between rural areas and the growing urban centres of western India. The Malis ranked as *Shudra* in the four-fold *varna* scheme.[1]

Describing the Mali caste of north India, the British ethnologist William Crooke stated that its 'primary occupation is gardening'.[2] He went on to declare: 'The Mali is a well-known figure in the folktales. The hero is often his son, or is protected by the gardener and his wife.'[3] Invoking the goddess Sitala of small pox, and acting as variolators and folk healers, members of the Mali caste associated themselves with the health and well-being of their area.[4] Many of them took on the surname 'Saini', an explanation of the term being that they were such admirable cultivators and gardeners, surpassed by none in industry and skill, that they came to be known as 'Rasaini', drawing from the word *rasai*, which literally means 'cleverness, skill'. In course of time, the first syllable was lost and the present name Saini left.[5] Denzil Ibbetson, ethnographer par excellence of Punjab, declared:

> The Mali, the *Malakara* or florist of the *Purans*, is generally a market or nursery gardener, and is most numerous in the vicinity of towns where manure is plentiful and there is a demand for his produce. He is perhaps the most skilful and industrious cultivator we possess, and does wonders with his land.[6]

The Saini community was later classified under 'Other Backward Classes' (OBC) in the Mandal Commission report,[7] therefore qualifying for government's affirmative action and reservation schemes in government jobs and educational institutes in the states of Uttar Pradesh, Punjab, Haryana, Rajasthan, and Madhya Pradesh. This essay underscores how Nek Chand

Saini gave new and creative meanings to designing a garden, as he collected and chronicled the planet, the earth, and its displaced people, providing them a language of dignity and hope amidst Anthropocene despair.

Nek Chand became a government employee in 1951 – a road inspector in the Department of Public Works (PWD), Chandigarh, at a time when the city was conceived as the capital of the post-Partition province of Punjab. The 'Chandigarh Capitol Project' was launched to showcase the modern development of an independent country, under the vision and guidance of the first Indian prime minister Jawaharlal Nehru, and the French architect Le Corbusier, along with his European team. Nek Chand has been portrayed as a great folk artist and architect, extraordinary collector and creator, sculptor and painter and has received numerous awards, including the Padma Shri in 1984 by the Indian government, in recognition of his significant contributions to the field of art.[8] Internationally, many individuals and organizations, art exhibitions and lecture tours, sculpture centres, and awards have focused on his life and works.[9]

The Rock Garden, spread over 25 acres of woodlands, is a magical miniature creation of a multi-layered, communal, and imaginative world of humans and non-humans, men and women, land and water, forests and mountains, village and city, ruler and people, war and peace, which also gives us glimpses of our planet and people. The garden was officially inaugurated in 1976, but there are no 'official' years of its construction and completion. As a master crafter, Nek Chand roamed for years in the Shivalik foothills, picking up stones and rocks that resembled birds, animals, humans, and multiple forms. He also relentlessly collected industrial, urban, and household waste on his bicycle, within the city and its peripheries, which was undergoing huge construction and engineering activities in the 1950s and 1960s. Within the thick forest, he secretly arranged those stones and rocks, boulders and broken chinaware, discarded fluorescent tubes and iron rods, broken and cast away glasses, windows and bangles, construction and steel wastes, coal and clay – all meant to give expression to his experience and understanding of the anthropogenic world, and its natural, human, industrial, and post-industrial life. Open-air sculptures and theatre trove, concealed gateways and common spaces, landscapes of mountain ranges and waterfalls, trees and forest patches, courtyards and chambers, streets and lanes – all ended and opened into a new mystic array that made 'Nek Chand's Rock Garden' an open-ended narrative of human history in India. Thousands of visitors – by one account 5,000 every day – come to visit this vast creation, making it one of the most popular public places in India.[10]

I see the story of the Rock Garden as one of Dalitbahujan (backward caste) Anthropocene and Nek Chand, in conversation with the place and the garden, chronicling his account of planet and people, and establishing a distinct everyday relation with waste. At the intersection of natural rocks and urban waste, Nek Chand presents an upfront local and low-caste view of human activities that capture changes in the local environment of Chandigarh, at a time of 'great acceleration' in a particular geography. Nek Chand's caste, caste consciousness, human labour, place, memory, and experience become important constitutive components in our understanding of multiple forms of Anthropocene that bear upon a society.

I attempt to provide a new narrative and framework of Anthropocene, by taking into consideration caste and human relationships established through the knowledge and lived experiences of Dalits and Bahujans[11] in nature. Scientists, geologists, humanists, and anthropologists have given certain meanings to Anthropocene, as the beginning of a new epoch. Since the past several decades, its origin and nomenclature, planetary and mega character, history of early evidence and warnings, overwhelmingly disastrous impact of humanity on earth's physical and biochemical systems, and agency of humans has been much discussed. Concerns and critiques of the Anthropocene as being undifferentiated, apolitical, Eurocentric, exonerating energy-intensive capitalism, and ignoring global inequalities has prompted counter proposals and terms like 'Capitalocene', 'Econocene', 'Petrocapitalist Anthropocene', 'Plantationocene', 'Homogenocene', and 'Plasticene'. Amidst these counterpoints, there have been arguments for the need to 'indigenise' and 'provincialise' the Anthropocene – connecting exploitation of the environment to exploitation of the indigenous and marginal segments of the human population, as well as building post-colonial, non-European chronologies, histories, and modes of land use and vegetation. Dipesh Chakrabarty, with reference to the earth scientist Jan Zalasiewicz, refers to 'many', 'different' Anthropocenes, where human-centred and planet-centred thinking are not entirely confined in the human timescales of history and politics.[12] Further, amidst different geopolitical contexts, Anthropocene and 'development' are considered coetaneous. It has been argued that implementation of development policies for the so-considered underdeveloped regions started to happen at the same time of what is known as a great acceleration of production, consumption, and environmental degradation at a global level.[13]

Taking cognizance of such valuable insights, I also build on ideas around Black-African Anthropocene, which consider the impact of race, racism, and racial capitalism on world ecology. Kathryn Yusoff describes Black Anthropocene as 'an inhuman proximity organized by historical geographies of extraction, grammars of geology, imperial global geographies, and contemporary environmental racism. It is predicated on the presumed absorbent qualities of black and brown bodies to take up the body burdens of exposure to toxicities and to buffer the violence of the earth'.[14] Narrating the social life of geology with a story of the 'Golden Spike', which contains blood and flesh of the human, non-human, and inhuman, he calls to unearth 'a billion Black Anthropocenes', who are, according to him, in 'experiential existence' and embody 'modes of mattering' that have no resource to agency of history. Gabrielle Hecht, critiquing the concept of Anthropocene, but also emphasizing its potential for political imagination, explores the notion of an African Anthropocene by focusing on creation and destruction of value/waste and past/future around a uranium mine in Mounana, Gabon, a county on the west coast of central Africa. She proposes three analytic moves to unwind the concept: different scale registers and their own narrative registers – spatial and temporal, local and regional, national and imperial; Anthropocene as an apotheosis of waste – 'the quantity, extent, and durability of discards', and how they create social relations; and 'putting the Anthropocene in place', where the story of a place becomes a reference point to understand the phenomenon and its multiple forms.[15]

I centre on Chandigarh city in Punjab state to explore Dalit and Dalitbahujan Anthropocene. Chandigarh, during the 1950s and 1960s, had enormous urban and economic activities, including huge construction, that changed the local ecosystem and generated a vast amount of waste. The vision and grounding of the city went through intense governmental, social, architectural, and technological processes. The place becomes a principal point to enter history and historical changes, where multiple issues – geography, displacement, destruction, construction, urban development, capital, market, and population – acquire prominent roles. I refer to place not only as a bearer and witness of accelerated human activities, a point of production of waste and change in local climate, but equally importantly as a core of alternative imaginations. By focusing on Nek Chand and the Rock Garden, I explain how in the same place and time, he not only built a critique of the Anthropocene through narrating his own story of the world but also showed immense human potential to create a different physical and social space for the present and the future. The Rock Garden, as a depository of waste, discards, natural, and human elements, records a complex trajectory of human history, where the discarded and devastated does not wither away from the environment in a few years, but needs to be revalued, recognized, and represented.

A brief account of the making of Chandigarh city acts as an entry point to the advent of political, economic, technological, governmental, and global acceleration in post-colonial India, initiated by diverse human actors. The works of Madhu Sarin[16] and Vikramaditya Prakash[17] have been especially valuable in providing historical details and ideological debates on Chandigarh.[18]

Chandigarh: Accelerating India into the Modern Age

Chandigarh – a city, a district, and a union territory – that serves as the joint capital of the two neighbouring states of Punjab and Haryana was the first planned city of independent India. Conceived in 1948, the city is characterized as showcasing 'the best experiments in urban planning and modern architecture in the twentieth century', a 'symbol of Indian modernisation', and a reflection of the 'changed environments, the industrial age, the new social order and the modern scientific technological advances'.[19] Built on a plateau on the foothills of the Shivaliks, which are part of the Himalayan range, Chandigarh was an entirely man-made environment, after clearing vast spreads of agricultural lands, villages, riverbeds, waterbodies, and mountain ranges. Since the conception of the city, there have been extensive research and publications on its architect and architecture, landscapes and population, buildings and monuments, city planning and administration. I am interested in how Chandigarh, in a particular historical period, signified mechanization, extraction, and domination of people, which radically changed the landscape of the region. The city was widely projected as a symbol of the new nation, industry, modernization, and machine age. Nation-building by the government acquired a new body and language of architecture, where urban planning and master plans became global projects, to spread fossil fuel and an energy-intensive economy.

The independence and partition of British India in 1947 created a historical setting for the origin of Chandigarh. The division of Punjab province into two, where the western half, with Lahore as its capital, was ceded to Pakistan, and the eastern half became a part of India, without a capital; Hindu–Muslim riots and violence leading to massive migration between the two borders; millions of homeless, suffering refugees, and the state government grappling with administrative, political, human, and psychological challenges loomed large over the region. Rather than establishing a new capital in an already existing town, or going along the Bengal way where East Bengal went to Pakistan without its historic capital Calcutta, who did not build a new capital and continued with Dacca, it was 'decided to build a new capital' of Punjab and 'locate it at Chandigarh'. Like its location, the name of the city was coined afresh: 'Chandi' represented the power (Shakti) of a Hindu goddess, which was the name of one of the then existing villages in the identified area, and 'ghar', meaning here the dwelling of the deity.

Amidst many complex rationales for the creation of Chandigarh, one was loud and clear – to remake the country, mechanically and technologically. The new government, institutions, and scientists invented, initiated, and imposed far-reaching interventions in the climate systems, which were also based on their promises to create a heaven on earth. The general resolve was that humans must seize the opportunity of freedom to march down the path of modernization, using their newly acquired power and autonomy. In times of such a push, it was not surprising that the making of Chandigarh city and the Bhakra Nagal dam, both seen as Siamese twins of modernity, were jointly propagated and developed as 'temples of modern India'. It was not a coincidence that both were also located not far from each other and architect Le Corbusier, who was authorized to build the Chandigarh city, was also invited to work on the Bhakra dam, and he even proposed a common symbol for both.

Nehru, India's first prime minister, was the key champion of the idea of Chandigarh, making it inseparable from the modern Indian state.[20] Not only did he openly express his masterly ideas and preferences about the city but he also practically involved himself in its everyday operations. His master's voice was readily adopted and amplified by the then chief minister, other ministers, and state machinery as a whole. Nehru's general belief in the greater good of science, technology, industry, and modernization had to be concretely sculptured into a new city. For him, the site chosen for Chandigarh was 'free from the existing encumbrances of old towns and old tradition'.[21] The new town was 'symbolic of the freedom of India, unfettered by the traditions of the past … an expression of the nation's faith in future'.[22] Nehru further elaborated his modernism at Chandigarh, in a context of new emerging architecture in the country, where a distaste for 'static' Indian mind and conditions was utmost audible:

> [The previous speaker] referred to the static condition in regard to architecture in India during the last two, three hundred years. That really was a reflex of the static condition of the Indian mind or Indian conditions. Everything was static –

there are bright individuals and bright movement but taken as a whole India was static....In fact, without being very accurate and precise, architecturally considered, for the last few hundred years, India was static.... Even before the British came, we had become static. In fact, the British came because we are static.[23]

Nehru brushed aside criticisms about the selection of the site, or the idea of the new city. He positioned his views on architecture, and was quite involved in the negotiations and selection of the architects. He continued his conversion with the architects during various stages of the building of the city. He intervened in the finalization of the masterplan. He suggested ways and means to deal with contested issues of land acquisition, land price, displacement, farmers' agitation, and unrests.[24] He had no hesitation in directing the then chief minister of the state regarding his preferred architect, 'I do hope that you will not overrule Corbusier. His opinion is of value.'[25]

Amidst a number of foreign architects, the renowned Swiss-French architect Le Corbusier and his European team were ultimately given the contract to shape Chandigarh. Corbusier had a much wider canvass on architecture. As a writer, painter, sculptor, and a key figure of Congres Internationaux d' Architecture Moderne (CIAM), which pioneered several new ideas for the Modern Movement in Architecture and Urbanism between the two world wars, he articulated some basic principles of modern town planning, like decongesting the centres of the cities, increasing the density of the centres of cities, increasing the means to circulate traffic, and increasing the areas of green and open spaces. He outlined the 'fundamental facts' of nature and the crucial importance of modern man and architecture-as-technology to be developed 'in harmony with the stars, the sun, the earth, the rivers, and other forces of nature'.[26] He visualized a 'Radiant City' – the city of tomorrow – 'of glass-and-steel skyscrapers set in parks; it would be the beautiful and efficient city restored to order and harmony through total administration of society by a great bureaucracy'.[27] At the same time, Corbusier and his team were candid and excited about the possibilities of modernity and machine, and how architecture and city planning should integrate itself with the demands of a new human age. They were jubilant to make alliance with industry and capital, and believed in human progress with technical advance. They had a firm belief in an 'elite meritocracy' – a collaborative of artists, technicians, and standard industrial production modes, taking a leading role in making cities, on behalf of the community.[28]

The coming of Corbusier in India has been critiqued from the perspective of decolonialization, and the nature of the post-colonial state. Gayatri Spivak characterized the gesture of Nehru inviting Corbusier to build Chandigarh on the pretext of decolonization as 'part of the script: the West on tap rather than on top'.[29] Corbusier came to India to translate Nehru's vision in architecture. Further, his own views on interface between nature, man, and machine created enough ground for his experiments in earth engineering – the largest of its kind in post-colonial India. According to Corbusier, a city was man's grip upon nature: 'It is a human operation directed against nature.... It is a creation.'[30] For him, machines were honest and efficient, thus their forms should be taken into nature. His ideal was an industrial society which should be 'centrally controlled, hierarchically organised, and administered from above'.[31] His model city was to be a 'City of Administration'. State authority was the key for the city's creation: 'Disillusioned by Western democracies, he looked to Stalin's Five-Year Plan, to Hitler's promise of full employment and to Mussolini's political posturings as proofs of future hope and authority.'[32]

Chandigarh ultimately came out as a unique, grand, world-known city – the scale of Capitol Complex on top of the town comprised the High Court, the Assembly, and the Secretariat; other monumental and symbolic buildings; the grid-iron V plans for motorized fast traffics; seven Vs; hierarchical sectors and residences; central plaza; shop-cum-offices and shop-cum-flats; land development; urban–rural divide with its immediate hinterland; division between formal and informal; controls on living population numbers and their dwellings within the city; strict parameters for cleanliness, order, and discipline; river beds, lakes, parks, and greenery as the 'leisure valley'; development of infrastructure, consumer goods and services – every idea and design was explicitly modern, exhibiting proudly the similarities with other prosperous capital cities of the world.

Nehru's dream of modernity and Corbusier's conception of 'second machine age' commenced in Chandigarh in a single decade, an era of gushing cement, steel, glass, stone, bricks, wood, gas, fire, and earth-changing construction work. Multiple forms of extraction – land and water engineering, civil and environmental landscaping, mining and drilling – were set into motion. Corbusier's qualified 'Reinforced Cement Concrete' (RCC), a modern-time 'artificial stone' with plastic qualities, became a prominent sign of the change. The city was characterized as 'a big massive concrete plate',[33] where structures were predominant. The vast 'V' network was created for petrodiesel-based mobility, embracing the power of an obsessive fossil fuel sector.[34] Corporate growth – spatially expanded and massively distributed – unleashed various economic and human activities, including capital investment, infrastructure,

trade and market, consumer goods and services, housing and buildings, informal
sector, labour, and unauthorized settlements.[35] The growth grid of Chandigarh
soon expanded across the city-state boundaries, and newly developed urban
areas like Mohali and Panchkula further accelerated industrial activities.
In the following years, vast expansion of satellite townships and defence-related
special projects, setting up of 'globally competitive' information technology
hubs, proposal for a 'an ultra-modern, futuristic city with global vision' for non-
resident Indians, massive acquisitions of land, and displacement of villagers
transformed the entire region into 'zones of hyperdevelopment'.[36] Alongside,
various acts like the Periphery Control Act 1952, which was further extended
and amended in 1962, was imposed on the project areas and its peripheries,
putting strict restrictions on poor people and their livelihoods. Millions of
displaced villagers; migrant, construction, and other informal sector workers;
and the urban poor were subsumed in the making of large-scale industrial
projects.[37] 'A labour force of men, women and children, underpaid, unhoused
and uncared for'[38] had to bear the destructiveness of the new industrial
economy and its catastrophic impact on them, as well as on the biodiverse web
of life in agriculture, mountain, rivers, and animals.

Chandigarh was a region-scale violence against places and species, as
well as against human beings. The coming of a new age of man and machine
particularly risked the lives of villagers, farmers, labourers, and people of low
castes. The speed with which the decision was made to build Chandigarh,
and with which all required human, technical, and scientific know-how was
acquired from the West,[39] needed the use of authority, force, and violence over
dissenting and protesting populations. The structures of democratic governance
were dismantled to facilitate an elite and exclusive decision-making. The site
selected for Chandigarh was an alluvial plain, gently sloping to the south,
right at the foothills of the Himalayas.[40] 'The region where Chandigarh was
to come up, it seems, was replete with deer moving around in groups of three,'
narrated a study.[41] The construction of the city meant immediate acquisition
of 28,000 acres of land, displacement of fifty-eight villages and their 21,000
living population, which had to happen under the centralist government and its
militarized police force, alongside extreme forms of extraction and inequality.
Thus, it was typical to propose a presidential rule on Chandigarh. Thereafter,
the Indian parliament enacted the Capital of Punjab (Development and
Regulation) Act in early 1952, to usurp central power in the Capital project.
This was followed by a series of acts, rules, and orders at central and state
levels to ensure a smooth passage to the new city. Such 'authoritarian edict was

established in accordance with general high modernist principles that dictated the suitability of science in designing utopian schemes for social life'.[42]

However, there emerged significant resistances for many years against the construction of the city, and to defend land and agriculture. In the project area, villagers, farmers, and labourers organized themselves to question a linear, mechanistic, and technological narrative of a place based on extracted land, minerals, energy, and labour. An 'Anti-Rajdhani Committee' was formed that organized protest demonstrations, *satyagraha*s, gheraoes, and hunger strikes throughout the state. People of various ideological streams – Gandhians, socialists, communists, Akalis, and even a section of Congress members came together to magnify the protest voices at the local level. They generally shared a common stake in anti-industrial political activism, based on a rejection of state-dominated environmental destruction. The government used the police force many times – arrests, jail, court cases, and intimidation – to suppress them. Local resistance had different phases – first, they tried to stop the acquisition and even stopped government officials from entering their land, then they asked for a better compensation, and later even went to the courts for legal intervention: 'Each phase was marked by an effort by the local people to urge the government to listen to their point of view and take a decision more in favour of local interests.'[43] Thousands of migrants, construction, and other informal workers were brought to the city, mostly through contractors. They received low wages without any social benefits and lived in temporary huts near the project sites. They formed a Capital Workers Union against the inhuman working and living conditions and demanded payment of minimum wages, implementation of labour legislations, and housing sites. Their organized, collective action was repressed by the state machinery. Eviction notices were issued to the construction workers to leave the city: 'The authorities ordered the forceful demolition of all the huts. Left with no other resources, the settlement residents prevented this by lying in front of the bulldozers, refusing to allow demolition until their demand for alternative accommodation had been met.'[44] Chandigarh even today continues with its concept of modern master plans and its authoritative norms.

At a provincial level, the making of Chandigarh represented a new modern age in India, the kind normally visualized in an era of colonialism and capitalism. In independent India, a local area had grown into a global playground in a single decade. The entire region lost all its land, as the area moved immense amount of soil and sediment annually, more than any other time in history. Its biodiversity was severely lost, and waste was generated all around. The physical landscape was totally transformed, in order to enter a modern industrial world of fossil fuel, automobile, transport, and service. The cement concrete created huge environmental challenges in terms of pollution, temperature, and humidity. In the enthusiasm of 'modern' acts that underline the making of Anthropocene, architecture and urban planning took a class and caste character, based on destructive, extractive, and exploitative exercises.

I now focus on Nek Chand, an ordinary low-caste person, who not only witnessed fast changes in his surroundings but also acquired a human agency to reflect and reorganize the ruins into a different physicality and materiality in the Chandigarh city.

Nek Chand: Life Narratives through Displacement, Discards, and History of the Earth

The life of Nek Chand has been represented richly, primarily through the Rock Garden. While the Rock Garden remains a key to understanding his story (which I will deal later), it also underlines long complex processes in which everyday lives and their reflections become a site for social reproduction of dominant economic and political order, but they also create grounds for resistance of that very dominance and hegemony. Life narratives have long been a constitutive archive and a performative mode of the oppressed.[45] Basing itself on different forms of life narratives – biographies, interviews, memoirs, art works, collections, and publications – this section seeks to identify some critical influences on Nek Chand's work. It reflects on how his everyday life journey was intertwined with the Anthropocene journey of the city, deriving from political, economic, technological, and architectural coalescences. My second move is to identify three major interrelated themes – displacement, discards, and rocks – as sites of indictment of violence in a particular place.

Developmental projects have been displacing millions of people every year from their home, land, and livelihood. There is no limit to human activities that can be labelled as 'development projects', nor are there real estimates about the number and extent of displaced people around the world. However, displacement of people is a key marker of Anthropocene and several streams of displacement studies have highlighted these interrelations. Studies on development-induced displacement, environment and climate refuges, climate change and migration, and mobility justice have come a long way in the past decades.[46] In such works, displacement is not only understood as forced migration of people and associated relief and resettlement but they also analyse complexities of displacement in relation to state, economy, society, culture, environment, climate change, national–international laws, justice, people's organizations, and protests. In this case, displacement enables us to understand intersections of multiple influences in and around Nek Chand. A master narrative of post-colonial construction of a 'city beautiful' dominated the imaginations about Chandigarh. However, displacement stories bring forth counter-narratives of lower, backward, poor, and dispossessed people. I also take references here from the diverse scholarship on waste, dirt, and discards. The generation of human and industrials wastes has also been characterized as an arrival of the Anthropocene epoch. In studies of wastes, economic and social relations, its entanglements with race, caste, and culture, issues surrounding durability, disposal and management, health, politics, and rights have gained prominence.[47] Newly emerged discard studies[48] have also opened the possibilities of seeing vibrancy and liveliness of waste, which can produce new cultural forms.[49] Nek Chand related to waste in innovative ways, to erase painful memories of the place, and instead create new human relations. Similarly, he collected rocks to narrate a history of earth and tumbled with them to demonstrate integral relationships between different biotic and abiotic components of the environment.

In his entire lifetime, Nek Chand lived with scars and experiences of displacement. His creation of the Rock Garden was a creative way to deal with the destruction and displacement all around him.[50] His childhood accounts, few and far, narrated his village life with family, land, agriculture, and animals, and how Partition destroyed it completely. In a climate of violence and disorder, Nek Chand set out from his native village with his parents, four brothers, and two sisters. For twenty-four days of their footloose journeys, they encountered many dangers and crossed the river Ravi:

> After trudging over sixty miles [100 kilometres], he finally reached the north of India and settled in the town of Gurdaspur with his family. In 1949, however, he left them to make his way to Karnal, where he joined the staff of the highway department. In the following years, Nek Chand was to live temporarily in Panipat and Faridabad.[51]

Both his parents died just after their exodus to India: 'Partition devoured them. Had there been no Partition I am sure they would have lived for many more years.'[52]

In Chandigarh, Nek Chand witnessed different facets of displacement and its discontents, where village, villagers, land, agriculture, and livelihood became regular targets of state force and violence. Discarded materials from destroyed villages, collected by Nek Chand, provided the initial impetus for his imagination of the Rock Garden. Threats of demolition also loomed large over the garden: when the garden was first discovered in the hidden wilderness of the city's northern edge, it was to be destroyed because of its 'illegality' and violation of the master plan:

> Early one morning in February 1973, an anti-malaria party came upon a stocky middle-aged man in his underwear arranging rocks and stones in the rubbish dump that was. And word flashed along bureaucratic channels up as high as the District Commissioner. Nek Chand wasn't sure whether he'd be congratulated or arrested – for destruction of Government (waste) materials (and unauthorised use of Government land). He wondered – he said – whether a bulldozer would rumble down and flatten everything.[53]

In the 1990s, the government decided to build a road right through the middle of the garden, for the exclusive use of VIPs. Thereafter, trees were cleared and bulldozers arrived at the garden wall to demolish certain structures.

The early experiences of displacement were overwhelming and earth-shaking in a post-colonial country, which resolved to their 'people' to ensure social, economic, and political justice. Chandigarh was one of the first places in India where multiple displacements unfolded: complete destruction of several villages, their land, and livelihood activities; duple displacement of Partition 'refugees' from their new settlements; villagers working as construction labourers in the developing city and living in labour colonies; use of the Land Acquisition Act of 1894; 'cheap' value of land and natural resources and denial of just compensation to the villagers; and state repression and people's resistance. The unquiet around displacement and dispossession simmered for decades and continued to haunt the city. The 'periphery' of the city was formally converted into 'a controlled area' to serve the 'centre'. The rural enclosure was to fulfil the daily needs of the Capitol. Le Corbusier wrote a letter to the Punjab Chief Engineer:

> My idea is that the farming must be conserved directly near (in contact) with the Capitol Area, so that the life expressed by the work of the peasant will be seen and appreciated: the order of the fields of sugarcane, wheat, etc. And little by little the way of modernisation of the peasant life will appear.[54]

There was thus a fetishizing of rural India's 'primitiveness', and furthering a paternalistic agenda of using modernism to civilize the non-Western world – 'an idealized image of rural life to provide sharp, side-by-side contrast with the progressive modernism of the Capitol Complex'.[55]

Living amidst such cycles of displacement, Nek Chand moved around the displaced villages and collected materials from destroyed households, employing 'squatter' techniques. Several of these are prominently displayed in the garden, an integral part of its sculpturing: broken pottery and ceramics, fragments of glass bangles, human hair, tube lights, and bicycle parts. According to a research study, this showed Nek Chand's desire to highlight and prominently display 'pre-Chandigarh village life'.[56] There were miniature villages created in the garden, with or without the materials from the displaced villages, but they had the reminisces of lost and crushed people. It was aptly observed:

> Chand has crafted a material record of the city's acts of construction and displacement. As a result, the Rock Garden is arguably the only place in the city that visually records the original settlements together with the developing and developed city, and that signals, however indirectly, the history of movement, construction, and both upwardly mobile and displaced lives.[57]

An important material fall out of the construction of Chandigarh was the generation of waste in construction, market, infrastructure, energy use, consumption, and disposability. In the 1950s, the Chandigarh building project was fast developing in ways by which agricultural soil, forest lands, and human habitats were turning into concrete masses of waste. Waste, like the displaced labourers, was a 'matter out of place',[58] signalling ruins of everyday lives, and traces left behind by the new-age machines. During this same period, Nek Chand, belonging to the Mali caste, often considered a part of wasted lives and refuses of society, became a collector and narrator of waste. Waste was not his 'birth-based polluted occupation', but it provided him ground material to sculpt stories of the past, present, and future of the earth. The collection of waste was not stigmatized hazardous work for him; it was a creative, challenging endeavour to constitute innumerable faces and facets of human–non-human worlds. Since the beginning of 1958, Nek Chand travelled to several sources of waste – individual, social, public, economic, and industrial – to comprehend their production cycles, and used them to build a new physical world. His son Anuj Saini states in an interview with me:

Nek Chand's caste was considered lowly and backward within the larger social circle, but he never allowed caste discrimination and subjugation to come in his way. He was dealing with wastes and discards not as a polluted and dirty person, but as an illuminated and creative person, who was trying to carve a world from below. In our childhood, he filled us with a sense of humanity and dignity, akin to being equal to everybody in terms of education, skills and opportunities, even in a caste-based place.[59]

Saini Samaj, an organization of people belonging to the Saini caste, considered a 'backward' caste in Punjab, facilitated Nek Chand in his lifetime.[60] Nek Chand kept in regular touch with the Samaj and participated in its social and cultural activities. However, 'Nek Chand was not limited to the Samaj. He had a complex and sensitive relationship with caste and community activities',[61] explained Harbhajan Singh Saini, vice president of the Saini Sewa Samaj in Punjab.

The Rock Garden was primarily constructed by waste materials of the city. Unlimited and unwanted, these discarded materials captured the eyes of Nek Chand, who started collecting and storing them regularly in a secluded forest area and began their casting and placing in the night hours. It is virtually impossible to make a list of his collections: broken crockery and sanitaryware, white pebbles, fused fluorescent tubes, huge limestone boulders, precast concrete, brick and cement tiles, sheet metal, woods, street light poles, red sandstone flints, tar drums, brick bats, terrazzo tiles, reeds, thatch, bamboo, rubble, coal slag, lime-kiln waste, cement, electric fitments, mud-and-hay plaster, foundry and engineering materials, broken electric plugs, blown-out light bulbs, steel slurry, iron alloys, different metals, machine and automobile parts, frames, mud-guards, forks, handlebars of bicycle, tree trunks and branches, steel bars, Coca-Cola bottle-tops, porcelain, damaged pots, scraps of cloths, wasted furniture and furnishings, chemical jars and containers, and tons of rubbles.[62] There are ample accounts of how Nek Chand collected the waste materials of the city. As a road inspector, he was well-versed with the new geography and economy of the city. He could thus correctly locate from where to collect what and how to hide them completely. He, on his old bicycle, roamed around construction and debris sites, and transported waste on its carrier. He went to displaced villagers, construction workers, and industrial localities, and asked people for their garbage. At his hidden place, he created a small, makeshift hut structure, like a temporary home of a displaced villager, and blocked its view from outside by arranging bitumen drums. He fired the bicycle tyres or abandoned cars to make lights in night at his workplace.

He used cement bags to cover his head, to protect himself from swarms of mosquitoes. 'No storm, thunder, rain, or even the fear of wild animals could deter him,'[63] recalled his son Anuj Saini, an artist and member of the Rock Garden Society. Later, he also developed centres at industrial, commercial, and colony areas to collect waste, and used small trucks to transport heavy materials.[64]

For Nek Chand, waste of the city was his key focus, and he concentrated his energy on the reconfiguration of wasted things. Simultaneously, as a collector, he was also gathering objects – rocks and stones – which held memories of lives and times gone by. On his bicycle, he travelled through the Shivalik hills, the seasonal Sukhna Cho, Patiala Rao and Ghaggar rivers, and gathered countless natural rocks and stones, feeling that 'they have a soul', and 'in every stone there is a human being'.[65] In their remarkable study, Bandyopadhyay and Jackson characterized Nek Chand's rock collection as a 'passion bordering the chaos of memories', and marked its various details and distinctions:

> Rocks were selected depending on their appearance, texture, erosion, difference from previously selected rocks and whether they resembled something else, i.e., a face, person or an animal.... In the early collection one can discern two categories of rocks: first, the rocks found on the riverbed – mostly basalt – eroded and shaped by its environment into intricate forms, clearly suggestive of humanoid or bestial formal qualities. Secondly, the heavier, smoother, more rounded and less eroded monoliths, often with deep penetrative incisions or apertures, suggestive of primitive amorphous life form, and thirdly, a small number of magma-like rocks evidently shaped by volcanic activity.[66]

The humanoid figures resembled celestial beings and dancers, conjoined bodies, eternal female and voluptuous female bodies, human faces, and many animal forms. They were of different sizes and colours, each one a distinctive composition and crystal structure, telling their life on earth, as well as demonstrating the result of changes that life has made to earth. They were neither chipped nor flaked, signifying their timeless, earthy, and archaic qualities. For Nek Chand, life and rocks co-evolved on the Chandigarh land: displacement and discards told stories of everyday human activities in the colonial, post-colonial, industrial, and modern eras of capital. At the same time, they were intertwined with changing times of geosphere and biosphere, crystalized in rocks and stones. The modern world of Chandigarh was one facilitated by concrete, cement, and minerals. However, rocks were providing geological, biological, and historical contexts of an evolving earth, which were equally important and interesting. Amidst the displaced, discarded, and rocky materials, Nek Chand remained a humble and simple human being, who had the will to transcend and transform his experiences. He observed:

> I regarded myself as neither an artist nor a craftsman. I myself was completely insignificant. I had no idea about anything, except for the fact that I was devoting my time to a task I was passionate about. I worked to the limits of my strength…. It all came from my heart and my imagination. My intention was to build a kingdom for gods and goddesses. It is a gift from God. This garden is more than an offering to God.[67]

Large-scale displacement and discards, destruction, and dislocation were radically altering places and people. Villagers, farmers, low-caste people, and the poor were paying the price for massive changes. Living amidst such wide-scale reorganization, Nek Chand tried hard to go beyond it, and created his small-scale alternative to explore his vision of place, time, and its substances, which was different from the modernist vision. Simultaneously, collecting rocks with multitude appearances, he tried to connect deeply with the evolutionary times of the planet and people, which provided him with new images and insights, to reflect on the present and future. A focus on the Rock Garden – Nek Chand's archives of Anthropocene – unravels his alternative imaginations.

Rock Garden: Kingdom of Gods and Goddesses, Humans and Non-humans

Located on the northern edge of Chandigarh, between Sukhna Lake and Sector-1 Capitol Complex, with its famed governmental buildings designed by Le Corbusier, the Rock Garden was, from the late 1950s to the early 1970s, an unknown forest site of wild and thorny bushes, where Nek Chand deposited waste for years and silently built his 'kingdom'. It resembled a multi-layered complex of earth and living systems; innumerable images of humans and animals; community of people, animals, and plants; situations of war and peace; and the beauty of human creation and transformation. In the 1970s, it was characterized as an 'illegal', 'unauthorised' structure after the government officials discovered the place, in the process of clearing and tidying up zones of thick undergrowth for the development of the city. The site became legalized and officially named as 'The Rock Garden' in 1976 after years of contestations between the creator Nek Chand, the state, and the inhabitants of the city, who came out on streets in support of the unique formation. From the 1980s, the Rock Garden continued officially, often in a tense relationship with the authorities, which has aggravated after the death of Nek Chand. 'Official apathy and discard, humiliation and pain, this is what we experienced in developing an alternative world, and now everything can be slowly destroyed by the commissions and omissions of authorities,'[68] said Harsh Kumar, a co-creator of the Rock Garden and the India trustee of Nek Chand Foundation.

The Rock Garden, in its materiality and visuality, is a curious mixture of multiple structural layers and uneven pathways, landscapes and lives, common open spaces, and hidden secret spiritual spots. To begin with, a wall with a frieze of birds – ducks, according to Nek Chand – surrounded the Rock Garden, protecting it from outside. The first or early phase of the place comprised natural minerals – river rocks, walls of pebbles and terracotta vessels, constructed waterfalls, and small, discrete huts, along with shades of planted trees and waterfalls, figures of women, and their colourful clothes. The world of rocks hosted a succession of Hindu shrines, squares, and chambers of empires, and was reminiscent of villages destroyed. The area created an impression of a unified single living organism – physical, biological, material, and spiritual. The living organism and its non-living surroundings formed a natural association, each with a distinct appearance. The passage to the second phase was subtle, where sculptures of figures, constructed out of collected industrial, urban, and domestic discards, were arranged in several small and big groups. Nek Chand regarded these hordes of human and animal figures as gods and goddesses of his imaginary kingdom. Immortal and divine creatures, labourers, old men and women, groups of schoolboys and girls, dancers, musicians, painters, players, police, soldiers, oxen, monkeys, elephants, dogs, lions, and birds appeared here in many sizes and moods. Made with metal substructures and wires, cracked glasses, ceramic pieces and bangles, bicycle seats, frames, forks, and mudguards, the waste materials created specific shapes, colours, and effects. Sacred entities from the Hindu religion also stood in between, depicted through life-size figures. The third, on-going phase, which was always 'a work-in-progress' for Nek Chand, opened into a huge common space with stalls, swings, and ornamentations. Lush, natural plants and huge artificial trees prominently captured squares. Together, these three phases, with recognizable and fantastical figures and interconnectedness between them, small-scale buildings, shelters, and complex structured dwellings in different, spread-out surfaces and textures, partitions and edgy walls, courtyards, cave-like structures and dark areas, stone walkways, narrower and longer pathways, waterbodies, rocks, stones, and plants create a deep impression of a living civilization.

The Rock Garden has been interpreted through its architecture, sculpture, and landscape.[69] It has been described as a 'spiritual garden', a place 'between Heaven and Earth, in celebration of primeval, telluric, and stellar covenants', and 'the path of initiation'.[70] For urban planners and architects, the Rock Garden signifies an informal approach to problems of planning, design, construction and art, development of indigenous technology, and low-cost self-housing experiments, in the context of Indian tradition and modernity.[71] It has been characterized as a counterpoint to Chandigarh: 'Le Corbusier's Ruin',[72] a place 'one is drawn in to touch and caress', with 'the colossal grave', 'the dignified ruin' that is Le Corbusier's Capitol.[73] According to Bonfitto, the Rock Garden 'might be thought of as a material archive of the region's geological and human history: the area's ancient fragments as well as its modern history of migration, displacement, and construction are inscribed, as it were, on the site's material components'.[74]

I see in Nek Chand's Rock Garden 'Allegories of the Anthropocene'. Elizabeth M. DeLoughrey describes allegory, literally 'other speaking', as polysemous, which emerges as a mode of colonial, political, and systemic critique through the use of irony, subversion, and multiple cultural creations. In times of climate change and Anthropocene, environmental discourse is rife with allegorical modes. This is not surprising because 'allegory is known for its embeddedness in history (time), its construction of a world system (space), and its signification practices in which the particular figures for the general and the local for the global'.[75] Allegory symbolizes a primary narrative record of the disjunctive relationship between the human and planetary environments. The central question is: what kinds of narratives help us navigate an ecological crisis, which is understood as local and planetary, historical and anticipatory? DeLoughrey analysed how indigenous and post-colonial island artists and writers expressed, through allegories of plantation, radiation, waste, ocean, and island, thoughts that encapsulated global climate change.

The Rock Garden is a public place of human-made sculptures in a physical universe of its own, creating a distinct sense of space, time, cosmology, and other forms of matter and energy. More specifically, it employs allegories of innumerable human–non-human figures and their myriad locations, made of a combination of waste, rocks, bricks, and rammed earth, to depict ecological changes in the region, dynamics of an eco-system, and agency of people and place.

The wastes of Chandigarh created the base for the Rock Garden. Ruins of modernity were spread everywhere. In every phase, the garden challenged the modern model of development as a narrative of progress. In a revitalized allegorical form, it weaved its own story of the earth, coined differently by Nek Chand as 'Kingdom of Gods and Goddesses', 'Animal Kingdom', 'Fantasy Kingdom', 'Sacred Kingdom', and 'People's Planet'.[76] There are several underlying allegorical tropes in the garden, which explore the interrelationship between Anthropocene, industry, and state that accelerate human and environmental catastrophe. Amidst such allegories, the Rock Garden also has a significant space for ecological and political engagements with people through their labour and cooperatives.

Sculptures had a representative value in the Rock Garden. Human, animal, and plant figures, in a variety of forms and expressions, were of central importance here, strikingly occupying big spaces. While diverse mass forms were added in different phases, waste, rock, and stone remained the primary materials. Time and again, they transformed from standing objects to community activities. For example, natural, village, and community ecosystems first appeared through clusters of individuals and small groups. Later, these were transformed into big clusters and spaces. These communities of organisms were not ad-hoc collections of individuals or populations; they represented an ordered, dynamic, and complex organization. Community attributes grew in form and structure: dominance of human and animal figures, species' diversity, mutual interrelationship among members of a community, and trophic structures. They represented life, community, love, joy, worship, and collective endeavours. Human and animal figures also continuously interacted with their surrounding environment to form a synergy with sacred, divine, and immortal figures. There were many devotional and spiritual images of Hindu gods and goddesses, created through a combination of sculptures, rocks, and stones, which sought to weave stories around mythical beings, creation, and life of spirits. There were also miniatures of displaced villages, which represented a disruption of the ecosystem, symbolized also in their dwarf sculptures. Mythical figures and creative narratives built around them almost vanished subsequently. In later stages, spatial variations occurred in sculpture structures, and they appeared few and far in physical numbers and temporal situations. The figures continue to be the key, but the natural, communitarian, and societal images prevalent in the initial phases have been largely replaced by images of despair, horror, deformation, anger, and individuality. Similarly, female sculptures offered a broader articulation of their lived and living life amidst urban and industrial acceleration. They appeared in significant numbers, many a time with bright, coloured glass bangles. Bangles, for Indian women, have aesthetic, social, personal, and political values. Subsequently, they appeared with broken glass bangles, which are considered a sign of serious tragedy and loss, ill-luck, and a forerunner of bad events. In several places, broken glass bangles became the prime material to carve out women figures, so much so that their heads and hairs were often amplified with small, crushed bangle pieces.

Broken materials were important elements for figurative sculptures and narrative themes in the Rock Garden. They were huge in numbers and varieties – bricks, wooden window frames, buckets, glasses, stones, rocks, building materials, electrical appliances, warehouse things, pits and puddles, rusty iron and scrap metals, doors, bottles, machine parts, asbestos sheets, concrete slabs, and broken furniture. For Nek Chand, nothing on the earth was ever lost and discarded. Every broken material had body and soul, dignity and worth, which continued to remain on earth. They must be recognized and placed properly.[77] Broken or crushed – meaning Dalit – has historically served as a key to highlight exploitation, coercion, and violence, which made up the life of the lowest people of Chandigarh. Nek Chand from the Dalitbahujan community turned the broken and the crushed into a creation – a place for counter-narration. Rather than working as an oppositional political space, the Rock Garden became a powerful counterforce for an alternative cultural and social imaginary of Dalit Anthropocene. It was no coincidence that many organizations and political initiatives emerged in the city and outside, around the Rock Garden. Despite their different location and emphasis, they focused their questioning on the development of Chandigarh, and on the power of alternative imaginations of the Rock Garden. These were often social and political mobilizations to resist state and corporate attempts to change the garden. They pushed the post-colonial state to at least live with such imaginations, thus adding new dimensions to questions of place, industry, waste, environment, and equality in development.

Bicycles can be located throughout the length and breadth of the Rock Garden. Every kind of discarded cycle parts – frame, rim, seat, spoke, tube, tyre, wheel, brake, chain, chainguard, cup, axie, handlebar, basket, bearing, fork, mudguard, brake cable, and mirror – were collected and converted into multiple human, plant, and animal figures. Anuj Saini described his father's fascination for cycle parts:

My father was a cycle man. In fact, Chandigarh initially began as a cycle city, and it was his own cycle that took him tirelessly to each corner of the city, and to villages, rivers, and hills for collecting materials. His unique ideas and dreams took concrete shapes through cycles. Take, for example, how in places, he has used the cycle handle turned upwards as the horn of animals. Or, the bears were created by him by turning the cycle frames upwards and decorating them with discarded black coals of the foundry. He utilized the chips used for flooring to make the white line on their necks. A little forward, you would come across some people dancing and celebrating, as part of a wedding procession. Some women are dancing – their ghagras made by turning the cycle's seating cushions upside down.[78]

Le Coubusier's new 'Machine Age' was based on a vision of vast expansion of economic, industrial, and administrative activities, which required to embrace the power of fossil fuel, automobile, and energy-intensive sectors. The spaces for roads and intersections were primarily designed for automobile traffic. The V7 system of roads – V-1 linking the regional roads that led to the city, V-2 representing the two major cross axial links to the city, V-3 for fast-moving traffics, V-4 for the slow-moving traffic going towards market, V-5 loop road and so on – was geared to encourage motorized people and their mobility: 'I remember the words of Corbusier who said that he planned the city – using automobiles as an implied module. It is not expressing itself in terms of volume. Cars. Cars. And more cars. Chandigarh has now one of the highest per-capita owners of automobiles.'[79] Bicycles, buried in the burgeoning automobile city culture, reappeared in the Rock Garden. Robert Stam has argued elsewhere that garbage signals 'the return of the repressed', which provides an opportunity to see social structures 'from below'.[80] Several images created with discarded cycle parts also made visible the violence and disappearance inflicted on the city.

The Rock Garden did not have a 'scientific', modern, architectural form. Its creator was a poor, lowly, government employee, who was born and brought up in a village. As opposed to Chandigarh city, Nek Chand adopted local ways for conceptualizing the landscape, for carving sculptures and art works, and in its construction methods. It was not a homogenous village or cluster of traditional villages, but an *anchal* (region) that found its expression in his creation. The *aanchalik* (regional) character of the region, filled with the local, the small, and the ordinary had been erased, but was now given a prominent place in the garden. In his comprehensive work, Bhatti demonstrated that Nek Chand's choice of spaces – change in space from narrow lanes to wide squares, use of *deori* to change and/or accentuate vistas, screens to build a sense of anticipation – represented what one frequently finds in a Punjabi village: 'Nek Chand's Rock Garden, like the Punjabi villages, is an apt example of the unity between landscapes and architecture. The site is selected and treated with such sensitivity that the garden actually seems to grow out of the landscape.'[81] Bhatti identified four areas – construction, architecture, art, and landscape – to locate the overwhelmingly regional elements in the garden. Weaving and embroidery, furniture and furnishings, wall paintings and bas-reliefs, had images of village deities. For example, there is the figure of Sanjhi Devi – Mother Earth – who is surrounded by the symbols of sun and moon, and vegetation appearing all around it. The Rock Garden did not demonstrate a desire to restore the old, traditional order, but created its own distinct blend of social and cultural symbols and traditions. Economy and society, industry and tradition, both erased memories of the poor and low-caste people. Nagaraj stated: 'The task of building a cultural memory is one of the first responses against this act of humiliation.'[82] The foregrounding of local, *anchal* and small should be seen in this context.

The Rock Garden does not follow any fixed ontology. Created by a person of the Mali caste, in an area characterized by displacement and destruction, it evolved over time, space, and scale, and sculptured complex manifestations of human, animal, plant, and life systems on earth. It can be characterized as a Dalitbahujan Anthropocene, as the wasted, discarded, and broken materials narrate a history of violence; they also sculpt new imaginations. The 'wasted' lives and materials come alive to recreate worlds, which are not bound by the economics and politics of the new machine and industrial age. The garden becomes a historical testimony and a humanist critique of the unleashing of accelerated industrial activities in a province.

In several interviews, Nek Chand talked of his rural, local, lower, non-scientific, and non-technical background, and the possibilities it presented for his creation, particularly in a place that had radically changed under the influence of state and capital. States Anuj Saini:

> It was fascinating to see how Nek Chand could easily and intimately connect with multiple things – river and rock, land and forest, waste and art, squatter and place. Beginning from a home that was first destroyed due to Partition, and then faced with threats of displacement in his new settlements, he could muster the labour, time, and energy to explore new physical and spiritual landscapes. This new space had no boundaries of environment, systems, and history, and its characters were endowed with mutuality, mobility, and freedom.

The Garden also looks like an extension of his borderless 'self' which also connected with the spiritual, the universal, and the mystical.[83]

Dipesh Chakrabarty sees such a Dalit's body 'as the body that makes us aware of all the networks of connections between different life-forms that enables humans, as a form of creaturely life, to survive. The Dalit's body is itself constructed nonanthropocentrically – it is always human with animals, live or dead, and embedded in the world of microbes (with its relationship to the handling of waste). In that sense, the Dalit's is what I might call the planetary body'.[84] Through his collections, sculptures, and imaginations, Nek Chand continuously and simultaneously moved into the past, present and future: to a past of earth's geology of human displacement, to a present display of waste and broken materials, and to a future of individual and community redemption through nature.

Nek Chand lived and worked in Chandigarh – a place and an epicentre of the new machine age and modernity in India. The landscape of the entire region was altered beyond recognition in a few decades, where the people and the environment had to live under a stringent imposition of laws and state agencies. In the garden thus, images of places and persons – villages and villagers, dwellings and localities – appear prominently. The standing sculptures of humans and animals, individually and in groups, draw attention to their existence, claim, and rights in a fast-changing physical world. The Rock Garden created an alternative landscape in which symbols of humans, non-humans, agriculture, forests, water, livestock, and biodiversity – discarded in modern development and alienated in a hostile environment – could find their identity and culture. Since the major dislocation of the lowly and the poor happened at the very beginning of development of Chandigarh, the sculptures of village, home, and the surrounding environment became prominent in Nek Chand's imagination. This took a new turn after a rapid increase in human industrial activities, when waste became paramount in the city. Imagined kingdoms, spiritual salvation, cultural symbols, and metaphysical lives were reinvented in the garden. Nek Chand's kingdom is a creative creation and a conscious imagination of 'myriad temporalities and spatialities and myriad intra-active entities-in-assemblages, including the more-than-human, other-than-human, inhuman, and human-as-humus'.[85] Such alternative cultures of survival, creativity, and resistance must 'collect up the trash of the Anthropocene, the exterminism of the Capitalocene, and chipping and shredding, and layering like a mad gardener, make a much hotter compost pile for still possible pasts, presents, and futures'.[86]

Discards defined Nek Chand, and he, in turn, redefined them radically. Nek Chand may be called an 'outsider', 'raw', or 'rough' artist. Yet, his artworks made of waste can be compared to the works of Western sculptors like Jean Tinguely, Arman, Jean Dubuffet, Gaston Chaissac, Auguste Forestier, and Andre Robillard, who applied refuse and waste materials in their compositions, and also produced new environments.[87] Nek Chand can also be characterized as a collector of the city's waste, and a narrator of violence of the state and the capital, generated in a particular place and time. DeLoughrey proposed 'the collector as allegorist' – 'a figure well suited for this temporal movement between past, present, and future. The collector assembles the ruins of uneven human history to provide new possibilities for meaning for our present and past as well as to "augur" the future, using allegorical modes as "interpreters of fate"'.[88] The collector Nek Chand gave a comprehensive account of the state of waste in post-colonial India, in a time of destruction of the landscape of 'Chandi', and alterations in human–non-human nature. Unlike pre-modern times, it was not in rivers and forests, rocks and mountains within which species were found. Instead, it was the overwhelming spread and visibility of waste – an anthropomorphized place – that captured the imagination of Nek Chand. As always, waste also created spaces of environmental casteism, by demarcating service classes, castes, and zones, and by marking peripheral and controlled areas. The lowly and the poor hardly figured in Chandigarh's framing, but Nek Chand's production, distribution, collection, and creation of waste made caste and class visible in the city. Nek Chand used all kinds of waste – industrial, domestic, individual, and social. He was not only interested in recycling and sculpturing it but was also positioning waste as a signifier of a restive society. In the vast spread of the Rock Garden, waste acquired temporal and scalar effects. It was like an 'Anthropocene rock', consisting of concrete, steel, and bitumen of the planet's cities and roads.[89] Nek Chand was also collecting waste from the displaced villages. The Rock Garden is thus filled with materials from such villages, and people who were left behind as 'wasted'. He establishes a historical conjuncture through his installations – moving backwards to waste and 'wasted' lives in history, to the waste produced in the contemporary era, to the imagination of a future of human–non-human earth, where one can find waste and wasted lives coming alive and celebrating life.

The Rock Garden is a popular public place, and it is also illustrative of a wider phenomenon of de-structuring the earth in a new industrial system. It came up in Chandigarh at a time of historic rupture in the region, and the application of a new scientific and technological knowledge. The rupture happened at multiple levels – natural and external environment, ecosystem, land, agriculture and energy use, and material and waste production. The experiences of the local, the poor, the lowly and the Dalitbahujans were different from discourses on modernity, development, and post-colonial nation-building at national and international levels. This was quite evident in several protests at local and regional levels, which were supressed. Nek Chand's rocks, sculptures, landscapes, figures, and narratives signified how radical changes impacted relationships between human and non-human, environment and society – physically, spatially, and temporally. His rocks were a testimony of past times of the earth. The wastes and broken materials narrated times of the present. His landscapes, characterized by a wide range of individual and collective figures and their entangled life stories, reflected his 'self' and the future. There is no future of humans, according to him, without a complex web of time, history, place, and species. His understanding of Anthropocene was certainly anchored in his time, but it also had an environmental imagination for the future.

The forthcoming essay will examine the role of Dalit and low-caste labour in both formal and informal industrial sectors in the country. Dalit labour not only represents the existence of individuals but also reflects the broader caste society. Therefore, it is crucial to investigate how labour is allocated across different sectors, the structural dynamics of the industrial environment, and how unjust and exploitative working conditions contribute to the employment and environmental crises, as well as trade union movements. Dalit labour faces numerous challenges, including caste-based discrimination within the workforce, antagonistic caste–class relationships, hazardous working conditions, the burden of industrial restructuring, labour redistribution, and engagement in trade unions. The outcomes of these struggles vary for them, contingent upon specific circumstances, particularly the level of organization and solidarity among themselves.

Notes

1. Rosalind O'Hanlon, *Caste, Conflict, and Ideology: Mahatma Jotirao Phule and Low-Caste Protest in Nineteenth Century Western India* (Cambridge: Cambridge University Press, 1985), 105.
2. William Crooke, *The Tribes and Castes of the North-Western Provinces and Oudh*, vol. 3 (Calcutta: Office of the Superintendent of Government Printing, 1896), 452.

   3. Ibid., 455.
   4. David Arnold, *Colonizing the Body: State Medicine and Epidemic Disease in Nineteenth-Century India* (Berkeley: University of California Press, 1993), 116–58.
   5. Crooke, *Tribes and Castes*, vol. 4, 256.
   6. Denzil Ibbetson, *Punjab Castes* (Lahore: Superintendent, Government Printing, 1916), 188.
   7. Along with Scheduled Castes and Scheduled Tribes, socially and educationally disadvantaged castes are classified as Other Backward Classes (OBCs) by the Government of India. The Mandal Commission (Socially and Educationally Backward Classes Commission) was established in 1979 by the Indian government to identify socially or educationally backward classes, and to consider the question of reservations for these people, to redress caste discrimination. The commission recommended reservations for OBCs in government and public sector undertakings, which was implemented by the government.
   8. There are a large number of publications. To name a few: M. S. Aulakh, *The Rock Garden* (Hyderabad: Tagore Publishers, 1986); John Moizels, *Raw Creation* (New York: Phaidon Press, 1996); Leslie Umberger, *Nek Chand: Healing Properties* (Wisconsin: John Michael Kohler Arts Center, 2000); Soumyen Bandyopadhyay and Iain Jackson, *The Collection, the Ruin and the Theatre: Architecture, Sculpture and Landscape in Nek Chand's Rock Garden* (Liverpool: Liverpool University Press, 2007); V. P. Mehta and Nek Chand Saini, *Rock Garden: A Vision of Creativity* (Chandigarh: Arun Publishing, 2010).
   9. The Nek Chand Foundation was formed in 1997 in the UK and the USA, with the aim of supporting Nek Chand's work and raising awareness about the Rock Garden throughout the world. The Foundation's website (https://nekchand.com/welcome, accessed on 3 February 2024) gives a detailed account of his international profile, exhibitions, and recognitions.
  10. I did fieldwork in Chandigarh during July–August 2021.
  11. A political term, 'Bahujan' means 'the majority'. It combines scheduled castes, scheduled tribes, and other backward classes, who are socially and economically marginalized by the Indian caste system.
  12. Dipesh Chakrabarty, *The Climate of History in a Planetary Age* (Chicago: The University of Chicago Press, 2021).
  13. Marina Dantas de Figueiredo, Fábio Freitas Schilling Marquesan, and José Miguel Imas, 'Anthropocene and "Development": Intertwined Trajectories since the Beginning of The Great Acceleration', *Revista de Administração Contemporânea – RAC* 24, no. 5/2 (2020): 400–13.

14. Kathryn Yusoff, *A Billion Black Anthropocenes or None* (Minneapolis: University of Minnesota Press, 2018), 5.

15. Gabrielle Hecht, 'Interscalar Vehicles for an African Anthropocene: On Waste, Temporality, and Violence', *Cultural Anthropology* 33, no. 1 (2018): 109–41.

16. Madhu Sarin, *Urban Planning in the Third World: The Chandigarh Experience* (New York: Routledge, 2019).

17. Vikramaditya Prakash, *Chandigarh's Le Corbusier: The Struggle for Modernity in Postcolonial India* (Ahmedabad: Mapin Publishing, 2002).

18. I would like to formally record my debt to their analysis.

19. There are several publications. For example, Sangeet Sharma, *The Corb's Capital: Journey through Chandigarh Architecture* (Chandigarh: A3 Foundation, 2014); Ravi Kalia, *Chandigarh: The Making of an Indian City* (Delhi: Oxford University Press, 1987).

20. Sharma, *The Corb's Capital*.

21. Quoted in Kalia, *Chandigarh*, 12.

22. L. R. Nair, ed., *Why Chandigarh?* (Simla: Punjab Government, 1950), 4.

23. Quoted in Prakash, *Chandigarh's Le Corbusier*, 9.

24. Kalia, *Chandigarh*.

25. Chandigarh Museum Archive, Chandigarh.

26. Prakash, *Chandigarh's Le Corbusier*, 16.

27. Kalia, *Chandigarh*, 73.

28. Sarin, *Urban Planning in the Third World*.

29. Gayatri Chakravorty Spivak, 'City, Country, Agency', *Future Anterior* 16, no. 2 (2019): 61.

30. Le Corbusier, *The City of To-Morrow and Its Planning* (Cambridge: MIT Press, 1971), 1.

31. Kalia, *Chandigarh*, 86.

32. Ibid., 86–87.

33. Sharma, *The Corb's Capital*, 74.

34. Prakash, *Chandigarh's Le Corbusier*.

35. Sarin, *Urban Planning in the Third World*.

36. Manish Chalana, 'Chandigarh: City and Periphery', *Journal of Planning History* 14, no. 1 (2015): 74.

37. Sarin, *Urban Planning in the Third World*.

38. Maxwell Fry, 'Chandigarh–New Capital City', *Architectural Record* (June 1955), 143.

39. Kalia, *Chandigarh*.

40. Prakash, *Chandigarh's Le Corbusier*.

41. Meeta Rajivlochan, Kavita Sharma, and Chitleen K. Sethi, *Chandigarh Lifescape: Brief Social History of a Planned City* (Chandigarh: Chandigarh Administration, 1999), 22.

42. Tracy Ann Buck Bonfitto, *The Rock Garden: A Study of Memory, Place-Making, and Community in Chandigarh, India* (Los Angeles: University of California, 2017), 106, https://escholarship.org/uc/item/80j5t8x4, accessed on 3 February 2024.

43. Rajivlochan et al., *Chandigarh Lifescape*, 24.

44. Sarin, *Urban Planning in the Third World*, 110.

45. S. Shankar and Charu Gupta, *Caste and Life Narratives* (Delhi: Primus Books, 2019).

46. There are several such studies. For example, Jean Drèze, Meera Samson, and Satyajit Singh, *The Dam and the Nation: Displacement and Resettlement in the Narmada Valley* (Delhi: Oxford University Press, 1997); Walter Fernandes, *Sixty Years of Development Induced Displacement in India: Impacts and the Search for Alternatives* (Delhi: Oxford University Press, 2008); Amitav Ghosh, *The Great Derangement: Climate Change and the Unthinkable* (Chicago: Chicago University Press, 2016); Irge Satiroglu and Narae Choi, *Development-Induced Displacement and Resettlement: New Perspectives on Persisting Problems* (London: Routledge, 2017); and Mimi Sheller, *Mobility Justice: The Politics of Movement in an Age of Extremes* (London: Verso, 2018).

47. I have referred to many studies in this area. For example, Mary Douglas, *Purity and Danger: An Analysis of the Concepts of Pollution and* Taboo (New York: Routledge, 1966); Gopal Guru, 'Archaeology of Untouchability', *Economic and Political Weekly* 44, no. 27 (2009): 49–56; Assa Doron and Robin Jeffrey, *Waste of a Nation: Garbage and Growth in India* (Cambridge: Harvard University Press, 2018).

48. For details, see Gay Hawkins, *The Ethics of Waste: How We Relate to Rubbish* (Lanham: Rowman and Littlefield, 2006); Joshua Reno, 'Waste and Waste Management', *Annual Review of Anthropology* 44, no. 1 (2015): 557–72; and Penelope Harvey, 'Waste Futures: Infrastructure and Political Experimentation in Southern Peru', *Ethnos* 82, no. 4 (2016): 672–89.

49. Jane Bennett, *Vibrant Matter: A Political Ecology of Things* (Durham: Duke University Press, 2010); and Myra J. Hird, 'Knowing Waste: Toward an Inhuman Epistemology', *Social Epistemology* 26, nos. 3–4 (2012): 453–69.

50. Bandyopadhyay and Jackson, *The Collection, the Ruin and the Theatre*.

51. Lucienne Peiry and Philippe Lespinasse, *Nek Chand's Outsider Art: The Rock Garden of Chandigarh* (Paris: Flammarion, 2005), 12.

52. John Moizels, 'Nek Chand: Creator of a Magical World', in *Vernacular Visionaries, International Outsider Art*, ed. Annie Carlano (New Haven: Yale University Press 2003), 67.
53. Moizels, 'Nek Chand', 32.
54. Le Corbusier, 'Chand Indes Urb Plan', General No. 11/30887 (Paris: Foundation Le Corbusier, 1953).
55. Chalana, 'Chandigarh', 68.
56. Iain Jackson, 'Politicised Territory: Nek Chand's Rock Garden in Chandigarh', *GBER* 2, no. 2 (2002): 51–68.
57. Bonfitto, *The Rock Garden*, 130.
58. Douglas, *Purity and Danger*.
59. Interview with Anuj Saini, son of Nek Chand, Chandigarh, 12–14 August 2021.
60. The Saini Samaj has hailed Nek Chand as one of their celebrated heroes. See http://www.sainionline.com/personalities/nek-chand-saini-creator-of-rock-garden, accessed on 1 September 2021.
61. Interview with Harbhajan Singh Saini, Chandigarh, 8 July 2021.
62. S. S. Bhatti, *Rock Garden in Chandigarh: A Critical Evaluation of the Work of Nek Chand* (Chandigarh: White Falcon Publishing, 2018).
63. Interview with Anuj Saini, Chandigarh, 12–14 August 2021.
64. Bhatti, *Rock Garden in Chandigarh*; Peiry and Lespinasse, *Nek Chand's Outsider Art*.
65. Bhatti, *Rock Garden in Chandigarh*, 218.
66. Bandyopadhyay and Jackson, *The Collection, the Ruin and the Theatre*, 24–27.
67. Quoted in Peiry and Lespinasse, *Nek Chand's Outsider Art*, 15.
68. Interview with Harsh Kumar, Chandigarh, 20 July 2021.
69. Bandyopadhyay and Jackson, *The Collection, the Ruin and the Theatre*.
70. Peiry and Lespinasse, *Nek Chand's Outsider Art*, 28.
71. Bhatti, *Rock Garden in Chandigarh*.
72. Vinayak Bharne, ed., *The Emerging Asian City: Concomitant Urbanities and Urbanisms* (New York: Routledge, 2013).
73. Prakash, *Chandigarh's Le Corbusier*, 71.
74. Bonfitto, *The Rock Garden*, 118.
75. Elizabeth M. DeLoughrey, *Allegories of Anthropocene* (Durham: Duke University Press, 2019), 5.
76. Anuj Saini, 'My Father's Kingdom of Gods and Goddesses', in *Chandigarh: An Anthology*, ed. Anuradha Uberoi (Chennai: Creative Workshop, 2021), 42–50.
77. Saini, 'My Father's Kingdom of Gods and Goddesses'.
78. Interview with Anuj Saini, Chandigarh, 12–14 August 2021.

79. Sharma, *The Corb's Capitol*, 74.

80. Robert Stam, 'Beyond Third Cinema: The Aesthetics of Hybridity', in *Rethinking Third Cinema*, ed. Wimal Dissanayake and Anthony Guneratne (New York: Routledge, 2003), 31–48.

81. Bhatti, *Rock Garden in Chandigarh*, 254.

82. D. R. Nagraj, *The Flaming Feet and Other Essays: The Dalit Movement in India* (Ranikhet: Permanent Black, 2010), 149.

83. Interview with Anuj Saini, Chandigarh, August 2021.

84. Chakrabarty, *The Climate of History in a Planetary Age*, 126.

85. Donna J. Haraway, *Staying with the Trouble: Making Kin in the Chthulucene* (Durham: Duke University Press, 2016), 57.

86. Ibid., 160.

87. Peiry and Lespinasse, *Nek Chand's Outsider Art*.

88. DeLoughrey, *Allegories of Anthropocene*, 100.

89. Jan Zalasiewicz, *The Earth After Us* (Oxford: Oxford University Press, 2008).

6

Industry

Caste of Labour, Dalits, Industrial Ecosystem, and Environmental Politics

Machinery and modern civilization are thus indispensable for emancipating man from leading the life of a brute, and for providing him with leisure and making a life of culture possible. A man who condemns machinery and modern civilization simply does not understand their purpose and the ultimate aim which human society must strive to achieve.... A society which does not believe in democracy may be indifferent to machinery and the civilization based upon it.

– B. R. Ambedkar[1]

In our days everything seems pregnant with its contrary. Machinery, gifted with the wonderful power of shortening and fructifying human labour, we behold starving and overworking it. The new-fangled sources of wealth, by some strange weird spell, are turned into sources of want. The victories of art seem bought by the loss of character. At the same pace that mankind masters nature, man seems to become enslaved to other men or to his infamy.

– Karl Marx and Frederick Engels[2]

Ambedkar and Marx expressed the aforementioned sentiments during the nineteenth and eighteenth centuries, respectively. However, even centuries later, these words resonate in the contemporary context of the Indian industrial world. The advent of machines during the Industrial Revolution held the promise of a positive transformation of labour and life on earth. It was seen as a potential solution to the pressing issues faced by Dalit labourers in caste-based occupations and economy. These issues included working in polluted occupations, unsafe working conditions, unhealthy environments, limited opportunities for progress, and degrading work conditions. Nevertheless, the combination of caste-based discrimination and capitalist exploitation of labour and nature has led to a series of conflicting and contradictory situations. The relationship between industries and Dalit labourers, as well as their impact on the environment, has faced intense scrutiny over the past two decades. A comprehensive study conducted on a burgeoning industrial area within the National Capital Region highlights both the challenges and successes experienced by Dalit labourers in their interactions with the realms of industry, caste, and nature.

The Wazirpur industrial area (WIA) in Delhi,[3] capital of India, is choked with small and medium enterprises, spread over a thousand plots. Metal, plastic, chemical, engineering, electrical, and transport equipment factories appear in a variety of shades – indoor and outdoor, gated and open, complexes and warehouses, tin shields and makeshift brick structures – where machines, labourers, raw materials, waste products, pumps, piping, drainage, cart, lorry, light, and sound come into sight from dawn to dusk, seven days a week. Two blocks – 'A' and 'B' – are well known for more than a thousand stainless steel utensils-producing units. Signboards are randomly displayed – as manufacturer, supplier, trader, exporter, wholesaler – at the factory front, also announcing their fine products. Stainless steel is an alloy – a combination of iron and other metals, containing 10–12 per cent of chromium that makes it different from other steels and 'stainless' all the way through. A range of kitchenware and cookware, tools, and other equipment made in stainless steel are in constant demand for their excellent polish, durability, hard-tough surface, and resistance to scratching and corrosion, whose attraction cannot be tarnished even after long use. Hundreds of steel workers move around in the area, without any specific labour markers. Their casual conversations are drowned by the hum of machines mounted on cement grounds lining the workplace.

Wazirpur has been a well-known industrial area of Delhi after independence Established as an industrial estate in the 1960s, it was an outcome of the involvement of existing governmental planning and resources into the process of industrial production.[4] At this stage, urbanization coincided with formal and informal industrialization and new, official industrial zones sprung up fast around the city. However, since the mid-1990s, the city's industrial landscape has changed massively, as environmental organizations and the judiciary have targeted the polluting industries of Delhi, including WIA, and recalled master plans to conserve the city's environment. Amidst several complex issues like closure-reallocation of industries and job losses, court judgements, Delhi master plans, environmental campaigns, and labour responses, I will focus on the socio-ecological structure of WIA and the evolution of stainless steel factories, in a specific context of the conditions of Dalit labour and industrial ecosystem, through a conception of labouring and labour process, and of alienation, by Marx in his *Capital*[5] and *The Economic and Philosophical Manuscripts of 1884.*[6] Together with Marx's understanding, Ambedkar's *Annihilation of Caste*[7] makes it possible to understand the unnatural division of labour and how it involves the subordination of man's natural powers and inclinations to the exigencies of the industrial rules, based on hierarchy. This is to bring together industry, labour, caste, Dalit, and environment in distinct ways. It shows how a stainless steel utensil, an everyday commended consumer product, was closely connected with labour, health, and environmental issues over work, life, and machine; how Dalits grapple with their caste identity in an industrial landscape; and how Dalit labourers, working in hazardous, risky, and uncertain economic and environmental situations, assessed their rights and built perspectives about ecological futures, through intense contestations over land, air, water, and place. I am not dealing with inter/intra-caste and gender relations, and the community and cultural manifestations of labourers in and outside the factories.

Marx observed that labour and the labouring process – work, objects on which work is performed, and instruments which facilitate the process of work, together called the means of production – is related with the past and the present, and to the degree of development of labour, and to the social relations under which work is performed. In the capitalist labour process, means of production is commanded and consumed by the capitalist, and maintained by the coercive forces of the state. Marx argues that in such a production system, a group, an institution, a society becomes alien at multiple levels: (*a*) to the results or products of its own activity, (*b*) to the nature in which it lives, (*c*) to other human beings, and (*d*) to itself. The basic essence of alienation is self-alienation, which also calls for a change towards the world of de-alienation. In the Wazirpur region, industrial and labour processes were characterized by the exploitation of labour and natural resources, and it was not only the workers who felt alienated but almost everyone whose life was dependent on the factory. As Marx said: 'Every self-alienation of man, from himself and from nature, appears in the relation which he postulates between other men and himself and nature.'[8]

However, the relations of hierarchy and domination, characteristic of caste society, were equally active in the 'subsumption of labour'. According to Ambedkar, caste society does not merely divide the labour. It also divides them into watertight compartments, where labourers are graded one above the other, on the basis of their birth and stratification of occupation.

Caste, as an active factor in the labour process, imposes dangerous roles upon labour, which also facilitate the imposition of 'dogma of predestination', and strict social discipline in the industries.[9]

Some notable historical and contemporary studies on the lower caste labouring people in Delhi and other cities demonstrate that the changing meanings of city, labour, environment, pollution, and activism have put considerable pressure on occupations that are seen as polluted, unclean, hazardous, and peripheral to the evolution of the city's urban core. A history of debates, strategies, and tactics over environmental concerns in Delhi shows important shifts over time. Different issues at different times – infrastructure and public health, nuisance and noxious trades, pollution and zoning standards and technoscience, and environmentalism through legal rights – have left their distinct imprint on how people dwell and labour in the city.[10] In colonial Mumbai, there was a close relationship between caste and stigmatized labour, seen in sanitizing the city, removing refuse, and collecting urban waste. This remained stable through periods of economic change and urbanization because dirt, as a cultural category, is part of a hereditary system that imprints physical and moral impurity on its actors.[11] A remarkable study locates the everyday life of low-caste Muslim Qureshi butchers of Delhi at the intersections of caste, class, stigma, market, and environmental discourse, where the butchers negotiate multiple challenges related to a transforming city and country.[12] The collective life of Dalit stone quarry labourers, located on the southern outskirts of Delhi, and working in an informal sector set-up beneath the radar of effective law and policy, has been weaved around three facets – stones, symbols, and sociation – which structure their common entity for everyday survival and struggle.[13] These studies provide significant insights regarding the intersections between Dalit, labour, environment, and urbanity, and a specific case study narrated later can expand their observations.

Steel utensil factories were located in many industrial areas – G. T. Karnal Road, Rajasthan Udyog Nagar, Samaypur Badli, Anand Parvat, and Okhla. However, Wazirpur had the largest number of steel factories. A majority of stainless steel workers were Dalits, migrated from Bihar, Uttar Pradesh, and Madhya Pradesh.[14] Over a span of twenty years, between 1995 and 2015, I visited the industrial area several times and acquired a mixed data set – fifty semi-structured interviews with male Dalit labourers aged between 20 and 65 years (stainless steel factories were an entirely male domain), and a dozen interviews with activists from labour and environmental organizations based in the city over the same time period. I also contacted factory owners' associations, local labour offices, and police stations while also witnessing labour activism and protest demonstrations first-hand. I spent a substantial time, mostly in the evenings, visiting, observing, and knowing labourers individually and collectively in their dwellings at Wazirpur village, Wazirpur J. J. Colony, Azad Colony, Udham Singh Park, and Sukhdev Nagar. Since factory owners did not allow me inside the premises to observe the production process, I decided to work in a factory for two days in 8-hourly shifts as a casual wage labourer – pretending to be a distant relative of a regular worker and in search of a job.

Let's see the structure and dynamics of Dalit labour in Wazirpur and how the changing industrial development and environmental situation since the 1990s exerted a significant influence on modes of organizing labour activities under the impact of economic uncertainties, social dislocation, and stratification.

Wazirpur and Dalit Labourers

In the 1960s, Dalit labourers entered Wazirpur factories mainly from the neighbouring states. They laboured hard in factories and tried to learn basic industrial skills. I met the second and third generation of Dalit labourers, originally from Uttar Pradesh and Haryana, who started working in the factories since the late 1960s. Some of them opened their own tea, grocery, and other small shops, enrolled children in schools, and bought permanent dwellings in slums and resettlement colonies. Pasha, a teashop owner in J. J. Colony, Wazirpur, and hailing from Hapur in Uttar Pradesh, had been a keen observer and participant in the establishment of factories, colonies, markets, transport, and infrastructure, where he could move from factory to the construction to his own teashop. Harish migrated in 1965 from rural Haryana and worked as a helper in a chemical factory. His wife Lata became a vegetable seller, and son Suraja private tutor to school children.[15] Dalit labourers got engaged in factory production and involved in urban life. Industrial areas represented a conglomerate of local formations – small settlements, regional clusters, social organizations – and continued spatial expansion and provided the labourers with selective mobility. The city represented an existence of choice and a reservoir of resources for material production. Utilization of existing potentialities and skills of migrant labourers also resulted in an on-going aggregation of their work zones. Still, the expansion of work life remained limited because they were constrained by economic resources, social stigma, and selective skills.

At this stage, Dalit labourers continued their active association with their villages, land, and agriculture. There also rapidly developed an informal system of recruitment of labourers from specific rural areas. For example, a significant number of labourers came from the Mirzapur and Banda regions of Uttar Pradesh, through their family and caste networks. Such penetration of rural relations into factory areas was also encouraged by the employers. Dalit labourers were concentrated mostly in the hazardous and dangerous processes of steel utensils and chemical factories.

The second stage in the development of the social and environmental structure of Wazirpur area was the fast, chaotic growth in the industrial activities of Delhi after the mid-1980s.[16] Steel utensil factories burgeoned, along with a growing concentration of chemical and metal industries in the entire Wazirpur area. The area witnessed a huge inflow of Dalit labourers from remote areas of Bihar, Uttar Pradesh, Jharkhand, Rajasthan, Punjab, and Haryana. The new recruits were mostly landless, agricultural labourers, who migrated in situations of acute distress. Signalling every aspect of economic and social domination of capital – exploitation of space and natural resources, violation of labour laws, health and safety issues, fragmentation and alienation of social life – Wazirpur turned into a 'dark zone'.[17] In the steel factories, labourers regularly faced violations of labour laws – non-payments of wages, lay-offs, accidents, and closures of factories were frequent. Dalit labourers were killed, injured, and maimed in industrial accidents, and suffered permanent physical losses in the dangerous production processes. The Delhi General Mazdoor Front, a trade union formed in 1980 in Wazirpur, repeatedly raised issues of various illegalities in these factories, and how they were ruining the labour and industrial ecosystem in collaboration with the government machinery. Labour and environment were under pressure not so much through pollution as through massive spatial transformations and technological hazards produced by the profit-seeking, messy factories. Summarizing the experiences of many steel factory workers who were laid off in late 1980s and early 1990s, 66-year-old Dalit Rana stated: 'We had to somehow survive to work in the changed industrial environment.'[18]

The third stage, which was stimulated by the decision of the Supreme Court in 1995 to close the polluting industries of Delhi and relocate them elsewhere (*M. C. Mehta vs Union of India & Ors*, Writ Petition No. 4677/1985), was marked by the deepening separation and antagonism between industry, labour, and environment. The court decision was followed by a series of interventions by the Delhi Pollution Control Committee, the National Green Tribunal, and the notification of Draft Master Plan of Delhi 2021.[19] This intensified the instability of the industrial area, deepened the exploitation of the labourers, and accelerated the disruption of their life, both in terms of its relation to the city and by way of its ecological structure. Dalit labourers recalled how a qualitatively new system of labour process and division appeared within the pores of closure and relocation. There were many complex, mutually interrelated processes at work. First was the increasing closure of factories, making labourers redundant. Second was the illegal and hidden operations of many factories, leaving labourers totally at the mercy of employers. Third was the formation of an industrial environment, within which the polluting presence of particular industries was reproduced through the 'polluting' Dalit labourers. This raised a new outburst against the steel utensil factories, as well as against their workers. 'It was the creation of a pollution syndrome. Dalit labourers were seen like criminals, as if they were primarily responsible for the closure and relocation of factories. We were here-there-nowhere,'[20] explained Dalit labour activist Awadh Kishore. However, for the steel workers, pollution prevailed because of thousands of old and dangerous machines; the environment was destroyed as a result of illegal and haphazard growth of factories; sanitation was under threat because of the poor conditions of working and living spaces.

I will now focus on the everyday working life of Dalit labour in stainless steel factories, in the 1990s and 2000s. The physical aspect of labour functioned in tandem, and was deeply entangled, with the caste question. Labour had a physical and a caste relation with machines and nature of work, which was expressed in the very process of production, and also found expression in their physical and social annihilation.

Life of the *Garam Rolla Mazdoor* in Wazirpur

Steel utensil workers were called *garam rolla mazdoor*, meaning hot iron worker. The factories were identified as *garam rolla bhatti*, meaning hot iron furnace. The labour process in the steel utensil factories was seen as an endless hot reproduction of iron and furnace. Labour's physical capacity and skills, raw materials, and the environment were perceived through the burning flames of the factory, in which the physical and social worlds of work and of Dalits were structured.

Considering the subject of labour, and a labourer's condition of existence, Marx placed the active elements – brain, aim, skills on the one hand, and natural, inanimate world, passive elements on the other. These reinforced each other. But to see how different human participants relate to one another in the labour process required consideration of the social relations within which that process occurred, explained Marx.[21] In a caste-ridden capitalist labour process, the capacity to work by labour and the value of labour determined by the capitalist were both mediated by relations of caste. Ambedkar characterized them as non-economic modes of domination and exploitation that fed into capital–labour relations. These non-economic modes also determine the deprived state of Dalits, where they lack capability, confidence, and dignity in social spheres.[22] Steel utensil workers were poor, mostly low-skilled Dalits, and through a complex process of social and local reproduction, they were concentrated in certain kinds of works. The factory owners, managers, supervisors, and contractors were non-Dalits, mostly upper castes from Delhi and neighbouring states. Workers' exploitation had many forms. The more general ones were expressed in terms of denial of wages and social benefits, while some were more precise in terms of caste discrimination and exclusion.

The workers recounted how they had to work twelve hours a day and their monthly wages were between INR 5,000 and INR 8,000. Overtime was rarely paid. Aside from the wages, there was no dearness allowance or any benefits – no holidays, house, or medical facilities. Factories were unregistered; there were no regular records of workers. In casual registers, workers' daily presence and overtime were marked. The contract system also operated, where workers were hired through contractors and were paid even lesser wages.

Dalit labourers were an essential part of the steel factory. However, they were always on temporary and contractual employment. Identified as low-skilled hot iron workers, they had less possibilities of getting work outside the steel factories. At the same time, they were not only separated spatially in the factory premises, but they also became the sole bearers of darkness, crowding, noise, pollution, and health hazards: 'We increasingly felt our Dalit status. Our factory, job, employment, place had remained merely an appendage to our lower status in society,'[23] said 40-year-old Ranvir, who was thrown out of the factory in 2001. Many Dalit labourers reflected on the cumulative effects of expansion of illegal steel factories in the 1980s and early 1990s, which were exacerbated by rapid qualitative degradation of the industrial environment. In many instances, expansion of steel industries, destruction of the environment, and increasing loss of life of Dalit labourers happened simultaneously in the 1980s and 1990s.

Touching on the objects of labour – nature and its physical components, raw materials, and products – I found the majority of steel utensil factories located in dirty, dark and dingy places, with little ventilation. The machines were old and poorly maintained. Steel workers worked in crowded, suffocating rooms, or sheds. The workplace was full of raw materials, dust, and smell. Workers dealt with hazardous steel plates and chemicals without safety measures. In my interviews, many workers cogently stated that the nature of objects of their labour was an index of the degree to which they had been categorized into a lower caste status. Working spaces were largely reserved and exclusive for Dalits, as neither were other factory people visiting there nor were non-Dalit workers joining to work with them. Dalits also compared their condition with other factories in the industrial zone, where they found better working situations. 'It was dirt all around in the industrial, natural, and social environments. We, unlike other labourers, were taken for granted to deal with certain objects. We were always identified with "untouchable" objects even in the steel factories.'[24]

Instruments of labour are considered a measure of human position in society's relations with the surrounding nature and are also a criterion of industrial progress. 'It is not the articles made, but how they are made, and by what instruments, that enables us to distinguish different economic epochs,'[25] said Marx. Instruments of labour include not only the 'bone and muscles of production' – mechanical instruments of labour, machinery, and tools; and not only the 'vascular system of production' – pipes and various capacities – but also the production premises, roads, canals, means of communication, and the land (as a factory site and the bearer of natural properties used in working on the object of labour), water, and electricity. Machines and other instruments in the steel utensil factories were the mediators and powerful intensifiers of Dalit labour's subjugated status, which also drastically transformed the industrial ecosystem in Wazirpur. Their operation was outdated, with no technological upgradation. The history of their development was at the same time the history of the steel industry's relation with the subjugation of Dalit labour and their poor health and environment.

The physicality and longevity of the production process was striking, where workers had to regularly use their hands, legs, and chests on the machines for long hours. First the big steel plates were shaped. Then the big pieces of steel were cut on the cutter machines. Small steel pieces were pressed on the machines to make them thirty inches long. After this, the pieces were put in an acid tank. Then these pieces were again taken to the machine, where they were further elongated to 65 inches long. This was followed by heating them on fire. These were again put in the acid tank. Then these pieces were put in the cold 'fork' machine and made 90 inches long. Then the process of moulding it into a utensil started. The 'circle' of the utensil was cut and put in the press machine. From it, the utensils came out. Then it was cut into the proper shape. It was twisted and moulded on the beading machine. Then it was polished and ready for the market.[26]

The dangers and hazards of working in the steel factories were a known industrial reality. However, it was also structured in relation to specific sets of technical, social, and cultural values and relations, where only Dalit bodies and capacities were found appropriate to use in certain ways. Dalit labouring bodies were marked as different and stigmatized, ideal to perform the most degrading labour. The physicality of the machine and that of labour were contingent upon caste-determined industrial norms. I found deaths, serious accidents, injuries, loss of body parts, illnesses, and diseases to be a regular and internalized phenomenon within the factories. Practically every Dalit household had workers with deep marks on their hands, legs, stomachs, backs, or other parts of the body. Workers explained the 'impossibility' of physically disentangling from the hazardous and the dangerous. Every steel worker had been a victim of minor or major accidents and injuries at some or the other stage. Workers talked about the deadly 'fork' machine that was always open and was used to press 30 inches pieces into 60 inches. While pressing the steel pieces on the machine, small particles escaped, and like a shot from the bullet, they could enter any part of the workers' body. Sometimes a worker lost his eyes, or his whole face was covered with blood, and if the piece entered his stomach or chest, the worker usually died.

It was precisely in the processes and instruments of labour that labour life was dehumanized. The machines further personified the cruelty of caste killing. It was also a process in which both the industry and the employer participated – by appropriating natural and human substances, and by alienating themselves from the lives surrounding them. Ambedkar saw this as a working of the caste system in labour and industry, where labourers were forced to take certain works. There is no freedom 'to change his occupation', 'not permitting readjustment of occupation', and 'individual sentiment, individual preference has no place in it'.[27]

I will now explain how the subsistence of Dalit labour was not only a reproduction of given relations between caste and society, but it was also closely aligned with the environment in which they worked and lived. Industrial zones initially came up in places of open and common lands, and expanded over the years, consuming a sizeable space, air, and water. However, the development of a new urban environmental intervention in the city from the mid-1990s[28] severely disrupted labour and its relations to society, as well as its ecological structure.

Interaction between Labour, Industry, and Environment

The development of steel utensil factories since the late 1980s was a specific kind of capital- and caste-based production activity. The factories were massively developed, with distinct forms of exploitation of human, natural, and social structures as a system of 'industrialising nature' and 'naturalising caste'. On the one hand, steel factories represented a condition for the functioning of production and capital, and its resources derived from the 'environment', and in that sense it was part of the environment. On the other, it was a system that reproduced itself in time and space through the appropriation of social stratification and natural resources. However, the factories not only over-consumed natural resources, disrupting their natural basis, but simultaneously also required a regular circulation of socially and physically subjugated forms of labour, for only then was it possible for the capital to further accumulate and reproduce. This dialectics between the natural and the material, the labour and the environment, the external and the internal provides a key to understand Wazirpur's socio-ecological structures after the 1990s. Further, the capital city after the 1990s represented a rupture from the natural–social node of capital ties and relations.

In the mid-1990s, the Supreme Court gave a definite character to such tendencies and produced deep fractures in the historical interrelations between labour, industry, and environment. The past was never ideal, but the present was earth-shattering. The court order (Application No. 22, Writ Petition Civil No. 4677 of 1985: *M. C. Mehta vs Union of India and Ors*) was not merely a judicial response to a petition but was also an image of the world in which labour and environment, industry and nature, industrial and natural environments were separated and juxtaposed. On a petition filed on the pollution of the river Ganga, the court decided to close the 'polluting industries' in non-confirming areas, who were in violation of the Delhi Master Plan, and they had to be relocated elsewhere. The court also directed to shift hundreds of 'noxious and hazardous industries' operating in the confirming areas. Even though outside the preview of the court decision, the owners of the steel factories in Wazirpur made this an excuse to close down the factories formally and started running them informally and illegally. Simultaneously, they made their case for an alternative site, with the purpose of acquiring industrial land. The working and living environment of Dalit labourers became more dangerous and hazardous. I met several Dalit steel labourers in the 2000s who felt straddled between formal and informal, legal and illegal with no certainty of their factory's existence and the future of their work. They merely heard about the master plan, pollution, and industries, but still continued as an appendage to the exploitative machine and capital, and were constantly subjected to hostile environments.

The master plans became major pointers for some kind of nature 'spaces', which could impede or contribute to the socio-ecological process of the city in a competitive struggle amongst sector, people, and occupation. The first Master Plan of Delhi in 1962 was projected as a guide for a 'new development' of the city, with zoning regulations, non-conforming industries, and their gradual shifting to earmarked industrial areas, land use restrictions, street and community facilities, and floor area ratio. Other plans followed. Scholars have seen the 'Master Plan for Delhi Perspective 2001' as a struggle over its spatial forces, where particular industries, labour, and work were categorically made out of bounds from the city.[29] In the name of 'an appropriate balance between spatial allocations for the distribution of housing, employment, social infrastructure, shopping centres, public and individual transport and so on and adequate arrangements and reservations to accommodate different kinds of physical infrastructure and physical utility systems', it recommended the closure and shifting of a large number of hazardous and noxious industries from Delhi within stipulated time periods (Delhi Master Plans 1961, 1981–2001, 2001, and 2021).[30] The court's decisions and subsequent actions by various governmental agencies qualitatively shifted the character of the city's environment. Dalit labour in certain industrial spheres was ousted from work and areas of their occupational concentration were changed, as were the distances over which they travelled. Delhi, which was earlier an open and messy carrier of systems of work, labour, leisure, migration, and mobility, was called for specific 'natural' prerequisites, bereft of polluting labouring populations. Dalit labourers gradually left Wazirpur to find work in new industrial areas, developed in the semi-urban areas of Haryana, Rajasthan, and Uttar Pradesh, in even more exploitative and unhealthy conditions.

After the Supreme Court decision in 1995–96, the closure and relocation of factories continued throughout the 2000s, including in cases of discharging effluents into the Yamuna river, and also violating residential and non-conforming zones. 'Polluting' industries were a point of convergence, referred to repeatedly by courts, government agencies, and environmental litigants. Images of Dalit 'polluting' lives, with their dangers and deaths, perceptions about their raw physicality and high toxicity, past links with certain production processes, and their social status in the industrial caste hierarchy served to further restructure linear and local stereotypes about the interrelations between labour, environment, caste, and pollution in the new-found urban environmental activism.

M. C. Mehta, environmentalist and lawyer, who brought the case to the Supreme Court, had his entire arguments based on the Master Plan and the polluting industries:

> It was a case of soft pedalling to industries. Government never implemented the Master Plan. These industries have polluted air, ground water sources, and, worst of all, the Yamuna. The problem of pollution is pathetic inside the factories also. Nobody, neither the industries nor the workers, wants to go out of Delhi.[31]

When I met J. R. Jindal, president of the Delhi Factory Owners Federation, to discuss the closure and illegal functioning of the steel utensil industries and their impact on Dalit labour, he expressed his inability to deal with labourers:

> There has been no socio-economic and technical survey of the industries in Delhi which could give specific details about the labour or the polluting industries. By a rough estimate, only a few hundred industries could be placed in the polluting categories. Steel utensil industries are largely small polluting units.[32]

The left trade unions were conscious of the fact that certain industries, like steel, chemical, and textile, were badly hit, and their labourers were left as most vulnerable. Yet, for example, the All-India Trade Union Congress, affiliated with the Communist Party of India, took a balancing stand that 'we are not for closure and relocation of industries. At the same time, we support the cause of environment and ecology. A proper balance is to be brought between environmental protection and development'.[33] After being asked about the Wazirpur steel utensil factories and Dalit labour, it was stated: 'Yes, special provisions could be made. The industries could be asked to change their nature of activities so that they are less polluting and dangerous. Enhanced compensation could be given to the suffering people.'[34]

As the city marched towards the closure and relocation of polluting industries, to provide cleaner air and water, labourers retracted into a bitter and complaining mood. Fearing the worst, Dalit labourers felt that the government, employer, market, and environment were together deserting them, leaving them with a bleak future. For some, the hard reality of sudden, unprecedented change was ruining their lives. For many, it was a hard truth that despite the claims of a cleaner environment, they continued with the same dirty and polluted lives. Nature perished together with thousands of abandoned labourers; it continued to be destroyed by illegal, unauthorized factories, running even more ruthlessly. It was vanishing as a result of the widening of pollution territories through court-directed and government-monitored upcoming of suburban polluting industrial zones. While the pollution- and environment-related concerns created different responses, depending upon people's social locations, people across social boundaries also felt that stainless steel utensils lost their usual sheen and shine.

Industrial Environment and Labour Struggles

A news report published widely in 2014 informed: 'Workers strike at Wazirpur Industrial Area.'[35] While there have been several strikes and workers' unrest at Wazirpur in the 2000s, this call witnessed a sizable participation of steel utensil workers. The strike raised some basic demands: enforcement of minimum wages, payment of overtime, provision of appointment letters, worker identity cards, salary slips, employees' state insurance (ESI), provident fund, bonus, safety measures, government holidays, and salary payment in the first week of every month. A workers' committee, Garam Rolla Mazdoor Ekta Samiti, organized the struggle. There were similar strikes in 2012–13, when workers achieved a monthly wage hike and weekly offs.

'Factory owners have registered a false FIR against the leaders', 'the police have only tried to suppress the strike by trying to disrupt workers' rally', reported civil rights organizations.[36] On 27 June 2014, after twenty-two days of strike, workers and employers signed an agreement in the presence of the Delhi Labour Commissioner. A release by the Workers Committee stated that the factory owners have accepted to implement the 8-hour workday, minimum wages, overtime, ESI, and provident fund. However, by 2 July, workers had to agitate again on the violation of the agreement.[37] A labour department official predicted that the increased minimum wages and social benefits for workers would make employers unsettled and such violent incidents would increase. The president of the Delhi Stainless Steel Trade Federation complained that the implementation of new wages and safety and environmental standards in steel and other factories was economically unsustainable. People in the workers' colony feared that police would come to intimidate and arrest them and that local landlords would ask for their eviction, to avoid trouble in the locality.

Life in the steel utensil factories was defined by labour, economic, social, and environmental troubles and uncertainties. Labour and their organizations, employers and their federation, locals, the police, and environmental authorities were fearful about the future. Their internal contradictions were acutely straining, where a course of action pursued by one end seemed to be generating a countervailing and opposed reaction by another. The history of the development of WIA, the nature of production in steel factories, and social relations involving labour and environment had a strong bearing on the labour and environmental conflicts since 1990s.

Everyday challenges of the industrial environment gave them an impetus to raise a wide range of traditional and socio-economic demands for regular employment and minimum wages, leave and payment of bonus, and new ones such as creation of a safe and healthy microclimate, introduction of new machines and technology, and reduction in work hours and health hazards. At the same time, each of these labour demands had caste and Dalit overtones, which emerged even more distinctly in the 2000s amidst particular environmental interventions. 'Pollution' became another area of sharp conflict between the labouring people and different social groups and strata of Delhi society. Labour struggles were convoluted and contradictory at times. On the one hand, Dalit workers were putting a great deal of effort to get the unions active on their economic, health, and environmental issues; on the other, they were gradually withdrawing from the industrial area, searching for new livelihood opportunities. They felt isolated due to the characterization of factories and their occupations as polluted, and in a state of social and ecological panic about an imminent livelihood and industrial disaster, at times opted out from the industrial spheres, withdrew from collective trade union activities, and instead chose the new informal sectors like private security, transport, construction, and service.

The organization and assertion of Dalit labourers went through different phases. As the labour force of steel factories was largely caste-determined, there was a mechanism for their self-organization. Such concentration of 'homogeneity' was at times critical for raising basic labour issues related to wages, payments, leave, and medical issues. It also made them realize the need for a certain level of collective correspondence between factory owners and steel workers. During the late 1980s, Dalit workers also started gravitating towards union activities. I met many labourers who joined the Delhi General Mazdoor Front, a left trade union, and raised economic demands. P. K. Shahi, General Secretary of the Front, recalled the history of organized struggle of Dalit steel workers:

> The Front originated in a steel factory in the early 1980s, when a worker died in an accident. There was no compensation paid to the worker's family, neither was anybody held responsible for his death. There was anger. We first decided to organize a demonstration of the steel workers outside the factory and they, for the first time, turned [out] in significant numbers. The agitation continued for some time, where steel workers of Wazirpur demonstrated against the local police, labour office, and labour commissioner. The Front managed to get compensation for the dead worker. The glaring caste character of the steel factories was a given fact. At the same time, everyday survival issues of Dalit labourers within the factory were extremely pressing, and demanded immediate attention of the union.[38]

Shahi narrated some unique characteristics of Wazirpur steel utensil factories. Most of the factories were owned by *bania* traders, belonging to a limited number of families and often in close relationship with one another, such as brother, uncle, nephew, or cousin. The system of contractors and sub-contractors was widely prevalent, and Dalit labourers were mostly employed through such systems. There was a close nexus between the factory owners, antisocial elements, contractors, and the police. The factory owners also had connections with people who commanded power in the local slums. Thus, workers' living places and spaces were under their indirect control. Labourers lived under constant fear and terror and thus avoided union activities.

Amidst fast and illegal expansion of steel utensil factories since the late 1980s, workers' health and safety issues became prominent, as the number of accidents and injuries increased manifold. Awadh Kishore, a prominent activist of the Front, told me that in the early 1990s, the union began to raise individual cases of injury or death in each factory by submitting memorandums and organizing demonstrations and sit-ins at the factory gates. In a few cases, workers could get leave, medical reimbursement, and compensation. However, in most cases, owners used to close down the factory and shift to another shed, with a new identity and a new set of workers. The labour department and the local police hardly intervened.[39]

Dalit labourers, in the course of unionization and individual struggles, demonstrated their willingness and courage to fight for basic labour rights.

More importantly, for the first time, the specific exploitative character of steel utensil factories of WIA came under the spotlight. The secluded and discriminated existence of a vast number of workers was no longer oblivious. Labour leader P. K. Shahi accepted candidly that it was not easy to gather wider support amongst the general workers towards the downgraded and stigmatized steel utensil labourers, as sectors like engineering, chemical, and transport were considered higher in the hierarchy of industries and labourers. According to Shahi:

> It was extremely important to create a joint front of the workers with other factories, organizations, and forums to strengthen the cause of Dalit steel workers. Steel workers were not able to achieve much gains out of the union's individual struggles. Factory owners and the entire factory system was extremely brutal towards them because of a combination of social, economic, and cultural factors. A Joint Action Committee was formed in 1990, which took up individual cases, as well as labour, health, and environmental issues of the sector as a whole.[40]

Dalits as factory workers acquired a voice of their own, and the Joint Action Committee could organize a series of activities – rallies, demonstrations, street plays, and songs – in which steel workers joined in large numbers along with other workers. A leaflet, circulated in one such rally, said:

> We all have to fight together to halt the death and exploitation of steel workers. Our struggle will have several dimensions – firstly, this is an issue to ensure relief for the injured and killed workers and their families. This is also an issue to stop such accidents and deaths in future. This is equally an issue to guarantee right to information and right to work in a healthy and safe environment. This is an issue to uphold right to safe life of common people.... We have a difficult struggle ahead against the combined onslaught of owners, corrupt police-administration and dealer unions. Only our struggle can ensure the right to life and dignity to steel workers.[41]

The initiation of a joint committee activated the central trade unions, who, while active in the industrial area, had not given much attention to the steel utensil sector. Lal Jhanda Union or Engineering Workers Lal Jhanda Union, affiliated with the Centre of Indian Trade Unions, though decided not to join the Joint Action Committee, submitted a memorandum to the Labour Commissioner focusing on the health and safety issues of steel workers, and organized demonstrations. Steel workers' issues in Wazirpur were slowly moving to the frontline of trade union struggles. In effect, the union activities, which focused on basic socio-economic needs of the labouring people, became critical for concerted efforts of trade unions and joint associations, both at the sectoral and the local levels, and held promise for anti-caste labour and environmental struggles.

The Wazirpur labour movement had entered a different phase after the Supreme Court's intervention on polluting industries. Importantly, the employment situation of workers was shaken, and the momentum of individual–collective struggles for the betterment of labour and industrial environment received a series of setbacks. Madav Prasad, a Dalit steel labourer, who was out of the factory since 2004, said that the new environmental concerns generated a thinking akin to 'environmental outcastes',[42] which advocated prevention of pollution at any cost, even by destroying industries and laying off workers. Proliferation of different groups on environmental issues and multiplicity of proposals by trade unions had also turned the situation rather confusing for Dalit labourers. For example, the Centre of Indian Trade Unions issued a joint statement in which they said: 'More than one hundred thousand workers find themselves face to face with loss of jobs.' Yet, they also said: 'Trade Unions firmly hold that protection of ecology and environment constitute an essential part of any sustainable industrial development.'[43] The Delhi General Mazdoor Front was of the view that closure and relocation of industries was not the solution to control pollution. However, they asked for 'alternative source of employment'. Civil society organizations were offering different interpretations of what must be done to solve the problem, practically and politically. An 'ecological liberalism' prevailed, where there was no stable alliance of organizations and the industrial situation was fraught with the dangers of disintegration. Dalit workers were feeling disillusioned with both labour and ecological movements. Such situations provoke Dalit labour 'to aversion, ill-will and the desire to evade', explained Ambedkar, when labourers are overall bound to a caste-determined economic organization and their hearts and minds are alienated from the work. Thus, 'the greatest evil in the industrial system is not so much poverty and the suffering that it involves as the fact that so many persons have callings which make no appeal to those who are engaged in them'.[44]

In the 2010s, Wazirpur witnessed the emergence of new labour organizations, like the Garam Rolla Mazdoor Ekta Samiti and Lok Adhikar Sangathan, who regrouped the workers in steel, chemical, and engineering industries together, and launched a series of struggles on their economic and health issues. The labour movement assumed a militant character and formulated a common action plan to unite workers in different sectors. At the same time, the caste and social character of labour problems also manifested in an upsurge of Dalit organizations in the area, like Dalit Ekta and Dalit Sangharsh Samiti. As Ram Prasad, Secretary, Dalit Ekta, who works in a steel factory and lives in Wazirpur J. J. Colony, said:

> Dalit workers need an organization of their own, as they cannot merge into one integrated stream anyway. For us, one of the important demands is the elimination of pollution and discrimination in every form, in and out of the factories, in both economic and social spheres. In seeking the fulfilment of these demands, Dalit workers comes out in defence of their life and dignity, not only in the factories but in the overall industrial environment.[45]

Dalit labour has long been understood in terms of caste-based traditional occupations, exploitation, degradation, and dignity. There should be exploration of Dalits in an industrial ecosystem, and how the industrial and environmental regimes construct their bodies and work inside the factories and the city. In the stainless steel utensil industry of Delhi, caste is embedded in capital, labour is embodied in the Dalit male labouring body and its physicality, and environmental and social pollution become the markers of industrial areas. Dalit labour crossed over local and regional boundaries, in the hope of building new futures for themselves. However, from their arrival in Wazirpur in the 1960s to their employment in 'A' and 'B' blocks, from the expansion of post-liberalization economic spheres to capital's growth and prosperity, from environmental activism against pollution to trade union movements, Dalit labouring bodies continued to be produced and reproduced, used and cursed, demarcated and disciplined in the industrial landscape of Delhi.

A complex, contradictory and layered narrative emerges around Dalit labour migrations to Delhi's industrial and social environments.[46] For many labourers across caste lines, the gradual transition from a rural to an urban industrial set-up in an early historical period grew and evolved into learning and practising new skills, along with an expansion of decent work opportunities and individual autonomy. However, in the post-1980s industrial scenario, steel labourers were entrapped amidst exploitation in the factory, insidious caste division of labour, and the often-fatal impact of deadly machines and a hazardous working environment. As the city galvanized to unexpected environmental concerns and activism, an emergent urban elite, drawn mainly from environmental, professional and civil sectors, tried successfully to tie pollution problems with implicit caste and class factors. Certain people and certain industries were seen as the main causes of pollution and environmental ills. Such exclusion was channelized through governmental agencies, as was the case in Wazirpur, and was selectively adopted by industries through increased informalization and illegality. There was a pattern where environmental, judicial, governmental, and industrial agencies worked together to define, regulate, and push pollution and 'polluting' labourers in the city – from stainless steel factories to other industrial areas of the National Capital Region. Notions of pollution and purity – a historical–social grid in Indian caste society for creating and treating the 'other' on the basis of birth, occupation, touch, place and dirt – penetrated into the city's imagination, to put the labouring people in general, and Dalits in particular, in their 'proper' place.

At the heart of labour, environmental, and health issues of steel workers was the problems faced by all labourers in an exploitative and hazardous working environment. The Wazirpur industrial area overall was becoming an exclusive exploitation zone. Dalit labourers and others continuously contested the repressive nature of their work and environment through individual and collective efforts – from joining the trade union ranks to looking for alternative occupations and locations. Yet, Dalit labour also moved in and out of the industrial environment and labour unions. Their politics of de-alienation – individual and social – found different forms and expressions, which culminated in the renewal of organized struggles, as well as in enhancing their human capacity for freedom, mobility, and creativity. Labour struggles thus put forth economic and social demands related to work, housing, rehabilitation, and livelihood. New trade unions and support groups emerged amongst the Dalit labourers and tried to expand spheres of collective interests.

The next essay will specifically concentrate on a vital aspect of the industry, namely technology, and its close connection with Dalit labour and the environment. Technology, as a tool of labour, serves as an indicator of the conditions experienced by the labouring individuals and their relationship with the surrounding nature and society. It also acts as a measure of historical progress. However, in numerous industrial sectors and occupational domains in India, technology becomes entangled with caste-based discrimination and hierarchical structures. The evolution and utilization of labour instruments in areas like the leather industry, both in the past and present, simultaneously represent the history of society's interactions with Dalits and nature.

Notes

1. Valerian Rodrigues, ed., *The Essential Writings of B. R. Ambedkar* (New Delhi: Oxford University Press, 2002), 157–59.
2. Karl Marx and Frederick Engels, *Selected Works*, vol. 1 (Moscow: Progress Publishers, 1978), 500.
3. Wazirpur industrial area, spread over 210 acres, was developed under the second Delhi Master Plan in 1966. It is owned by the Delhi Development Authority (DDA) and operates under the Municipal Corporation of Delhi (MCD).
4. Surajit Chakravarty and Rohit Negi, eds., *Space, Planning and Everyday Contestations in Delhi* (Delhi: Springer India, 2016).
5. Karl Marx, *Capital*, vol. 1 (Moscow: Progress Publishers, 1975).
6. Karl Marx, *The Economic and Philosophical Manuscripts of 1844* (Amherst: Prometheus, 1988 [1927]).
7. B. R. Ambedkar, 'Annihilation of Caste', in *The Essential Writings of B. R. Ambedkar*, ed. Valerian Rodrigues (New Delhi: Oxford University Press, 2002 [1936]).
8. Tom Bottomore, *A Dictionary of Marxist Thought* (New Delhi: Oxford University Press, 1983), 12.
9. Ambedkar, *Annihilation of Caste*, 263–64.
10. Awadhendra Sharan, *In the City out of Place: Nuisance, Pollution and Dwelling in Delhi c. 1850–2000* (New Delhi: Oxford University Press, 2014).
11. Shireen Mirza, 'Figure of the Halalkhore: Caste and Stigmatised Labour in Colonial Bombay', *Economic and Political Weekly* 53, no. 31 (August 2018): 79–85.
12. Zarin Ahmad, *Delhi's Meatscapes: Muslim Butchers in a Transforming Mega-City* (New Delhi: Oxford University Press, 2018).
13. Mohammad Talib, *Writing Labour: Stone Quarry Workers in Delhi* (New Delhi: Oxford University Press, 2010).
14. A majority of the workers came from Chamar, Dusadh, Musahar, Jatav, Ahirwar, Dhanuk, Kori, Pasi, Balmiki, and Khatik castes. Some were from other backward classes and castes like Yadav, Kurmi, Bind, Teli, and Kalwar. Very few were from forward Brahmin or Rajput castes.
15. Interview with Pasha, Wazirpur, Delhi, April 2000.
16. Ram Singh Bora, 'Migrant Informal Workers: A Study of Delhi and Satellite Towns', *Modern Economy* 5, no. 5 (2014): 562–79.
17. Mukul, 'Steel Workers of Delhi: At the Mercy of the Owner', *Forum Gazette* (1989): 8–9.
18. Interview with Rana, Wazirpur, Delhi, January 2001.

19. The National Green Tribunal and Delhi Pollution Control Committee have given several orders against the steel utensil factories in Wazirpur. Draft Master Plan of Delhi 2021 also included them in the 'negative' list of industries.
20. Interview with Awadh Kishore, Shalimar Bagh, Delhi, March 2008.
21. Bottomore, *A Dictionary of Marxist Thought*, 267.
22. Rodrigues, *The Essential Writings of B. R. Ambedkar*, 38–9.
23. Interview with Ranvir, Wazirpur, Delhi, February 2002.
24. Interview with Rakesh Bahadur, Wazirpur J. J. Colony, Delhi, March 2001.
25. Marx, *Capital*, 175.
26. Mukul, *Steel Bartan Udhog: Maut Se Jooghte Mazdoor* (Delhi: Delhi General Mazdoor Front, 1989), 18.
27. B. R. Ambedkar, 'Philosophy of Hinduism', in *Dr Babasaheb Ambedkar: Writings and Speeches*, ed. Vasant Moon, vol. 3 (Mumbai: Government of Maharashtra, 1987), 68.
28. Gautam Bhan, *In the Public's Interest: Eviction, Citizenship, and Inequality in Contemporary Delhi* (Athens: The University of Georgia Press, 2016).
29. Stephen Legg, *Spaces of Colonialism: Delhi's Urban Governmentalities* (Oxford: Blackwell, 2007).
30. Delhi Master Plans, retrieved from https://dda.gov.in/about-master-plan, accessed on 25 September 2021.
31. Interview with M. C. Mehta, Delhi, October 1996.
32. Interview with J. R. Jindal, Delhi, October 1996.
33. Interview with D. L. Sachdev, Delhi, October 1996.
34. Ibid.
35. *Indian Express*, Delhi, 16 June 2014.
36. Peoples' Union for Democratic Rights, Delhi, 22 June 2014.
37. Workers Committee, 'Memorandum', Delhi, 3 July 2014.
38. Interview with P. K. Shahi, Delhi, March 2014.
39. Interview with Awadh Kishore, Delhi, March 2008.
40. Interview with P. K. Shahi, Delhi, March 2014.
41. Struggle Committee of Labour and Mass Organisations, 'Labourers of Wazirpur Industrial Area – Unite and Fight!' (Delhi, 2001).
42. Interview with Madav Prasad, Wazirpur, Delhi, January 2015.
43. Coordination Committee of Trade Unions, 'Press Release' (Delhi, 2016).
44. Ambedkar, *Philosophy of Hinduism*, 68.
45. Interview with Ram Prasad, Wazirpur, Delhi, 2015.
46. Eesha Kunduri, 'Between *Khet* (Field) and Factory, *Gaanv* (Village) and *Sheher* (City): Caste, Gender and the (Re)shaping of Migrant Identities in Urban India', *Samaj* 19 (2018), https://doi.org/10.4000/samaj.4582, accessed on 3 February 2024.

7

Technology

Tanneries, Tanners, and Technological Injustices

Technology plays a crucial role in mediating the interaction between society, labour, and nature. In a broader sense, the history of technology reflects the evolving attitudes of humans towards labour and nature. The dominant socio-cultural factors in a society significantly influence the development of technology, technological sciences, and education, and the strategies employed in transforming the natural world. One important principle that shapes technological application and innovation across various industrial sectors in India is the casteist materialist conception of the relationship between labour, work, and nature. This conception perceives this relationship as a natural, static, and fixed phenomenon, rather than a dynamic and evolving process. According to this view, the interaction between labour and nature is shaped by caste-based cultural and historical processes, with its specific content and course determined by the hereditary occupation and skill of Dalits. In practice, the combination of caste domination and capital constantly seeks to control and regulate the exchange of technology and knowledge between labour and the physical environment, which ultimately proves detrimental to both Dalits and the environment. This approach hampers progress and perpetuates inequalities, hindering the positive impact of technology on society, labour, and nature as a whole. The leather industry in Kanpur, along with the associated technology and the age-old involvement of Dalit labour, serves as a prominent example that illustrates the aforementioned scenario.

Jajmau, an industrial suburb of Kanpur in Uttar Pradesh, is well known for leather tanneries. Located on the banks of the river Ganga and administered under the Kanpur metropolitan area, its tanneries have been doing good business since long – generating INR 15,000 crore of revenue every year and accounting for the country's 30 per cent of leather export. Tanning is the process of treating skins and hides of animals to produce leather, and tanneries are the places where the skins are processed. Tannery workers are Dalits – mainly Chamars – and low-caste or low-skill Muslims. Kanpur, founded in the early eighteenth century, became an important commercial and military hub in British India and after independence, it came to be known as the industrial and financial capital of Uttar Pradesh, and a producer of fine quality leather and textile products. The tanneries in the city increased rapidly between the 1900s and the 2000s. There were hundreds of tanneries and leather factories, and millions of formal–informal labourers who were directly or indirectly employed in the sector. The tanneries' technological system – factory, machine, design, work, labour, education – did not change much, particularly in terms of its relation to caste, and continued to reflect a caste culture. The entanglements of caste with technology reflect that the caste system has permeated our technological processes. Historically, the leather industry played an important role in the formation of caste identity by embedding Dalits – physically, materially, and visibly – into a polluted occupation. Simultaneously, the endurance of tannery technological systems in Kanpur and elsewhere in the country, in spite of several changes in industry, market, and city, demonstrated its deep socio-economic roots.

This essay engages with issues of caste, technology, and Dalits through a case study of tanneries in Kanpur. Technological systems underpin major aspects of the Dalit experience in nature and also promise to enhance the status and quality of their life through changes in degraded and dangerous work. The essay demonstrates how casteism has continued and strengthened through tannery technology, how technological injustices are hampering Dalits, and how the politics of caste and environmental justice is linked with a transformation of technology, particularly in caste-segregated occupations. Dalit tannery labourers have participated in various initiatives to ensure technological justice, so that issues of caste stigma and occupational health and safety can be addressed. The pollution of the river Ganga, the government's Ganga action plan, and its impact on the tannery are not dealt with in this essay. Hindutva politics of animals in north India, closure of slaughterhouses, and shortage of raw materials to tanneries in UP have also not been covered. Both have been significant political developments in the past two decades, leading to temporary or permanent closure of several tanneries in Kanpur and loss of livelihood of tanners. At the same time, they bear little relation to the technological systems of the factories.

Environmental justice movements have dealt with science and technology in a variety of ways. Struggles against environmental racism and injustices focused on industrial and technological systems that were producing health and environmental hazards for black and poor communities. Simultaneously, they demanded radical changes in discriminatory industrial and technological practices. Questioning the dominant practices of industry and science, they also searched for alternatives. It has been observed that

> science and technical practice are constantly in a process of transformation, that transformations in the direction of a more environmentally just science and technology are possible and are fostered by the engagement of technical practitioners with the EJ movement, and that transformative projects are made difficult by heterogeneous structural constraints.[1]

This has bearings on caste, Dalits, and environmental justice.

Technology, Caste, and Dalits

Technology and society shape one another, and there are intrinsic connections between technological experiences and social relationships. The term 'technology' has been defined comprehensively and critically, and includes material, physical, social, cultural, political, and economic aspects that encompass artefacts, work, labour, structure, tools, processes, management, organization, social and economic relations, money, power, and authority.[2] Technology is characterized as 'inherently ideological', which can drive society to reorganize itself to accommodate it. It has been stated: 'If our technology has been created mainly by the capitalist system, is it not probable that it bears the marks of its origin, a technology for the few rather than the masses, a technology of exploitation, a technology that is class orientated, undemocratic, inhuman, and also unecological and nonconservationist?'[3] Technology–society relationships have thus been explained through the prism of different ideologies and socio-economic systems of pre- and post-colonialism, imperialism, capitalism, and socialism.[4] Proposals for small, decentralized, appropriate, human, environmentally sustainable technologies have also been extensively discussed.[5] Feminists have critiqued science and technology, and emphasized it exclusionary social and political construction, as well as its male-dominated power, authority, and values that are embedded within the social processes of scientific and technological innovation.[6] They have also presented alternative paradigms to conceptualize and practice science and technology through a gendered lens.[7] African Americans have analysed the relation between race and technology, where issues of technological access and divide in American history, obsolescence of Black technological knowledge, and struggles towards 'transformative access to technologies' have been highlighted.[8] It has been argued that racial ideology deeply influenced technological prowess and it became a means by which people of colour were subordinated throughout the country. American whiteness and its technological capability were established as natural and dominant.[9]

The caste system has been explained as 'hierarchically graded, locally integrated, occupationally and ritually specialised, having endogamous social strata'.[10] This exploitative system has to be supported and sustained by several social organizations for its continuity. The physical and material structures of work and labour – process, organization, roles, rules – are forms of social organization, which are constructed specifically to maintain caste order. Machine and technology function in their specific arenas, but they are designed and maintained in a manner that reproduces caste hierarchy. Scholars have analysed the dominant role of caste in many dimensions of technology, including its access, diffusion, and innovation. 'India's tech sector has a caste problem,'[11] and several reports have detailed how this sector, from the tech companies to the technical institutions, are dominated by *savarnas*. Low-caste people and Dalits have to undergo everyday humiliation and marginalization in these places.[12] Technology access has been exclusionary and is based on caste. The promise of technology to make possible work with dignity, and to transform dirty, defiled, damaged labour and working conditions into humane and sustainable occupation has been belied because of the wide prevalence of a caste-based system, where low-caste, untouchable labour has not been considered worthy to even live humanly. Caste order can structure the experience and deployment of technology.[13] States Bezwada Wilson:

> Caste is deeply linked with sanitation. Dalit community cleans septic tanks and sewage. It is a shame that our country, which is focusing on the Mars mission and satellites, does not have resources to come up with proper machines and technology to clean drains and manholes without humans entering them. Those on the margins, in particular Dalits, face continuous problems of meaningful innovation and access to humane technological systems.[14]

A study on the relationship between digital communication technologies and Dalits in peri-urban Bangalore demonstrated the durability of caste in peri-urban metropolitan India, the social construction of usage of information communication technologies (ICTs), and myopia in the conventional understanding of a digital divide in India.[15] Technology diffusion is found to be impacted by a stronger social network within caste than across caste categories, thus limiting or expanding its potential.[16]

Technological systems are often identified with machines, tools, and the overall design. However, the history of design has been exclusively based on caste and class. Designs can play a critical role in enforcing and maintaining casteism. In an important study on tank irrigation in Karnataka, Esha Shah, anthropologist and historian of science and technology, showed how tank designs, an important component of irrigation technology, have been coded with dominant interests of caste and class that structure water distribution in a certain fashion and maintain social order. The task of creating and maintaining social order was delegated to technological design. And the social order around water distribution was reproduced through a reproduction of designs.[17] Another study on tank irrigation technology of Tamil Nadu in south India found that the instrumental dimensions of technology are inseparable from its social and political meanings, as they express class relations, caste status, prestige, and honour.[18]

Dalits, at the bottom of caste hierarchy, have been victims of technology injustices in many ways. Dangerous and unsustainable technologies have often functioned through their physical labour. Technological innovations have bypassed Dalits, and their access to existing technological systems have remained limited. Sustainable and innovative technological systems are seen as a great enabler for Dalits to break caste-based discrimination. According to B. R. Ambedkar, 'Machinery and modern civilization are thus indispensable for emancipating man from leading the life of a brute, and for providing him with leisure and making a life of culture possible. The man who condemns machinery and modern civilization simply does not understand their purpose and the ultimate aim which human society must strive to achieve.'[19] Given the nature of scientific and technical progress, Ambedkar was equally emphatic in his understanding that the transformation of nature by powerful economic and technological forces was not only having a living impact on separate components of landscapes but was also closely associated with the possibility of changing society altogether, and with it, its inseparable biosocial organ, that is, human and humanity.

In Ambedkar's world, technology and machine become only a beginning point for Dalits, in their whole movement from dispossession to power. In order to reach it, the 'untouchable' has to go through an arduous journey of acquiring knowledge, tools, and human resources, to produce something for him and for society. There are some crucial moments in this interaction of Ambedkar's man, machine, and technology: (*a*) untouchables' identity, since the work is at the same time their signature of introduction, freedom, and movement; (*b*) their intermediation, since one is accomplished by the touch of the other; (*c*) their individual and collective transformation, since work generates fulfilment of basic human needs. In the end, a machine–man–technology chain, in a given social and natural environment, is a life-changing process, which can, in theory and practice, be unlimited. Nature, through man and machine, not only serves to satisfy existing wants but also brings new ones into being.[20]

Many other anti-caste thinkers and activists envisaged a world where technological systems freed Dalit labour from their bonded and degrading work. E. V. Ramasamy Naicker, popularly called 'Periyar', and one of the chief progenitors of the Dravidian movement in south India, viewed that in a society organized around *varna dharma*, labour possessed no inherent worth or dignity. Elaborating on 'caste worker' and 'wage worker', he found it extremely demeaning when one had to labour particular tasks and not any other, and when the privileged refused to work and insisted on living off the fruits of others' labour. Hard and tireless labour got subsumed in traditional and political rhetoric. Extremely critical of caste occupations, Periyar argued for machine, science, and technology to eradicate caste divisions in labour. Addressing himself to artisans, Periyar stated:

> With the advent of the machine in all spheres of manufacture and production, they needed to enhance their skills. If they failed to mechanize their production processes, they would be left behind to suffer in outmoded and inefficient forms of production. Besides, technology, apart from enhancing the workman's skills, would also help him overcome the drudgery, tedium and exhaustion that were often the lot of many an artisan. Technology would also make available to him a certain amount of leisure and workmen could utilize this to better their life in ever so many ways.[21]

Anti-caste and Dalit thinkers have pointed out that exclusion and inaccessibility is only one part of larger technological injustices that operate in a caste society. Dalit experiences, memories, innovations, and contributions have been erased from dominant technological environments. However, 'productive knowledge systems of Dalit-Bahujan communities' have creatively provided day-to-day scientific technological processes during work. Dalits have also been characterized as 'subaltern scientists', 'social doctors', 'unknown engineers', and 'food producers'.[22] The historic contribution of Dalit agricultural techniques and knowledge in the rice fields of Kuttanad region in Kerala has been extensively documented.[23] Technology use of Dalits, along with its own images, languages, and experiences, needs to be taken seriously.

This discussion suggests that caste, technology, social order, and authority over Dalits are mutually constituted. Dalit lives exist in a dual state of technology existence, where access, inclusion, innovation in technology, and recognition of Dalits in technological systems has the potential to positively change their lives. However, the ground situation denies such a potential. In the following section, I see how particular areas and occupations give a distinctive caste to work, interweaving caste, technology, and Dalits.

Kanpur, Leather, Labour, and Environment

The history and geography of Kanpur played an important role in determining which industries and occupations came to dominate over others. Amidst a diversity of economic forms and cultural values within the broad canvass of colonial and post-colonial developments, it was the unique location of Kanpur – characterized by river, rural landscape, low-caste working population, and accessibility – that made possible tanneries and the leather industry to come up and grow in the region. The British Empire and its army used the location and specificity of the place to promote their political, industrial, and commercial activities. Kanpur was described as a creation of the colonial rule, growing from the site of a military camp into a town, whose history was inextricably tied to British presence.[24]

Kanpur city, formerly Cawnpore, in southwest-central Uttar Pradesh, is situated in the lower Ganges–Yamuna Doab, which is quite close to the Ganga river as well as only 72 kilometres away from Lucknow. According to the researcher Syed Asif Ali, the 'urban geography of Kanpur' first and foremost indicates the importance of geographical factors, which were directly or indirectly favourable to the development of an urban concentration, and evolution of a townscape and industries since the eighteenth century:

> The river Ganga was the most notable advantage of the site. Besides offering adequate water supply and a good protection from hostile invasion at least from one side, it was the principal means of the transportation for thousands of years in the region and during the early British rule, there was a great volume of water borne traffic between lower and upper regions of the Ganga Valley. A large share of this traffic was eventually passing near Kanpur because of its midway position on the Ganga, and this helped very much to develop Kanpur as an important

waterway junction, a factor of great importance for the economic prosperity of
the town.[25]

The centre of tanneries in contemporary times, Jajmau village, was historically
located on the riverbank, which also intersected with ravines for many miles.
Most industrial areas developed above the stream because of easy availability
of water, as well as protection from the floods. The riverside strip was the
main reason for the East India Company to establish a cantonment here in
1778, which also served as a military conduit for safety and the expansion
of economic and political activities of the empire. The vast region occupied
by the alluvium plain of the Ganga made the neighbouring rural areas
agriculturally productive, with a high density of human and domestic animal
population. The entire region was known for the production of wheat, rice,
cotton, sugarcane, hides, and timber.

The city came up as a corollary to the increasing penetration of the empire,
and the British army provided the initial impetus for the establishment of
textile, leather, and other industries. The growth of industries and labouring
population has also been influenced by complex changes in rural, caste,
nationalist, and labour politics in the region. Labour historian Chitra Joshi,
in her remarkable work on the Kanpur working class population and its
intertwining with colonial rule, military, state, community, nationalism,
and class struggle, has chronicled how the industrial city was constituted
and reconstituted in different historical periods. Long before the factory
industries, Kanpur city and cantonment had settlements of populations of
artisans, leather workers, weavers, tailors, and other miscellaneous labourers.
With the establishment of the cantonment and the expansion of military, civil
and economic activities, the army occupied long stretches of the river front
and displaced the old residents, who inhabited the riverfront in the Jaimau
area. They were pushed back, away from the river.

The 1857 revolt and its aftermath significantly changed the city and
the ruined cantonment areas witnessed the emergence of new industries,
including leather factories like Copper Allen and Company. According to
Joshi, it was a politics of space and sanitation that crystallized divisions and
boundaries between different areas and works. Leather factories and workers,
artisan and coolie labour, scavengers, porters, carpenters, and blacksmiths – all
kinds of labouring poor – were pushed into separate areas in the course of the
city's and industrial development: 'By the early years of the twentieth century,
factory chimneys dominated the city and the politics of sanitation shaped a
landscape in which squalor and working class bastis became synonymous.'[26]

Leather tanneries and labourers emerged as one of the main carriers of caste-based occupations and segregation in society. They became a central site where labour, technology, and Dalits came together closely. The city's tanneries and leather establishments increasingly organized work for its Dalit labourers in a manner that required only hereditary 'dirty' skills. Treating skins and hides of animals has long been an important element in the formation of caste identity in India. Its new organization in the city – exclusive location, concentration, factory, and technology – was buttressed through networks of rural and urban, working site and home, process and product, factory and river. Tanneries and technologies evoked different sets of caste materiality, the legacy of which was hard to shake even in modern times.

The expansion of leatherworks in Kanpur happened along with several developments and divisions among the Hindus of UP around caste function and occupation. Historians have richly analysed the dynamics of low-caste and 'untouchable' people vis-à-vis employment, culture, politics, education, health, and identity. Charu Gupta showed how the urban economy of UP, including Kanpur, saw considerable industrial expansion in the early nineteenth century. New jobs were created in relatively respectable occupations, with lower castes being appointed in railways, as manual servants of British families, as peons in offices, as municipal sweepers and scavengers. There was increasing migration, a weakening of hereditary employment, a loosening of traditional caste ties, some extension of leisure time, and a simultaneous forging of new alliances, which gave people a limited sense of liberation and security. There were examples of acquisition of wealth and status by members of inferior castes like Chamars, Doms, Telis, and Kalwars on account of the development of leather, oil-seed, and metal businesses.[27] Nandini Gooptu mentioned that in the 1920s and 1930s, 'untouchable' caste groups from rural areas migrated to Allahabad, Benares, Kanpur, and Lucknow. However, they were absorbed almost entirely in ill-paid, menial service jobs or in work connected with handling leather, in keeping with their 'traditional', 'low', or 'impure' occupations. There was both occupational and residential segregation, which tended to ossify the low status and social subordination of the urban untouchables.[28]

More particularly, an increasing demand for leatherworkers in Kanpur city and elsewhere led the Hindu society and the colonial administration to construct an occupational image of Chamars that suited them. Their characterization, presentation, and treatment in governmental and non-governmental spheres served to stigmatize and enrol them into leather works. They were represented as impure, criminal, and inferior, with a menial capacity.

Such crude representations encompassed their bodies, tools, and technological competence. These perceptions were further crystallized in designs of tanneries and working conditions of Chamars in these industries. Ramnarayan Rawat, in his research on Chamars and Dalit histories of north India, explained how they were deliberately encouraged to become labourers in the leather industries in the first half of the twentieth century. In fact, the majority of Chamars had been skilled cultivators and agriculturists in the state and only a small number of them were involved in some kind of leatherwork. This fact was turned upside down by Hindu and colonial constructions of them. They were treated like a criminal caste and accusations of cattle poisoning, and cases, including arrests, were rampant against them. Rawat states:

> The 'natural' association of Chamars with leatherwork was created during colonial times. Dominant representations of the leather industry, the opening of the leatherworking schools specifically for Chamars, and the colonial association of Chamars with the 'centuries-old *jajmani* system' all helped to ensure that leatherwork became their sole occupation and identity in the eyes of others. And this happened regardless of whether Chamars had ever actually been involved in full-time leatherwork.[29]

In a similar manner, leather-related science, technical education, and skill-building also essentialized Chamars with leatherwork. It has been stated that the leather chemist, trained to impart scientific knowledge to leather workers, often failed to negotiate the caste-based sensorial nature of leatherwork, thereby allowing caste to limit the reach of modern science in the industry.[30] Technical education in leather production in India between 1900 and 1950 got interlocked with intense structural and ideological hierarchies of caste and its conjoint markers of stigma, untouchability, and smell:

> The curricula and programmes ensured that outcaste workers who laboured not just on the shop floors but also completing experiments as 'coolies' in the laboratories and research tanneries could never move beyond the stigmatising work they did. Work processes and wage grades peculiarly became categories of birth.[31]

In post-independence India, the Kanpur leather industry, including tanneries, demonstrated the dynamics of integrating into national and global production and export networks. The micro, small, and medium-sized enterprises (MSME) and local lead and satellite firms got integrated with Dalit-based tanneries and technology.[32] There was a reconstruction of the caste and capital power relationship in a globally integrated sector, in which tanneries and Dalit labourers were often overlooked. Studies have pointed out that because of the industry's close links to social structures, new organizational structures that emerged often contained elements of continuity with traditional structures, as well as some that represented a break with them.[33] Kanpur tanneries developed in many ways – rural, small, organized, and finishing units. With the active support of the government and the private sector, hundreds of tanneries flourished in the region. They worked to their full capacity, without any modernization in their technological system.[34] Most of the Kanpur tanneries depended on a relocation and reorganization of caste labour and related power relations in local production networks. The growth and export enthusiasm of the leather industry rested on a 'Chamarisation' of leatherwork.

Environment and caste were thus intrinsically interwoven into tanneries and technology by Hindu society, the colonial administration, and post-colonial regimes. This interrelationship between caste and technology was further facilitated by processes of militarism, colonialism, industrialization, and markets in the nineteenth and twentieth centuries. It permeated to state ideologies and policies. This is what Kosambi meant when he stated that the 'adoption of technology of Hindu society' impacted 'the organisation of production'.[35] In the following section, I unpack the relationship between caste and technology by taking a look at the world of Jajmau tanneries. After all, it is the work of Dalit labourers in the tanneries that created leather products and contributed much to India's industrial growth.

Tanning, Tanneries, Technological Systems

Jajmau is located in south-east Kanpur, and on the southern bank of the river Ganga, which is also close to the military cantonment area. It was outside the main city in earlier times and thus became an appropriate ground for upcoming tanneries. The location was perceived as making city dwellers free from dirt, stench, waste, and high consumption of water. The area today is filled with tanneries and tannery labourers. There are dilapidated and crumped houses of labourers, smoke coming out from tannery chimneys, dust in the air, slush, watery mud, leather waste, and remains of hides on streets and narrow lanes without drainage, little open public spaces, and crowds of men, women, and children all around. Tannery buildings are closed from inside and researchers are normally not allowed to go in them. Interviews with labourers, field observations, and literature on the leather industry gave me some clues about their technological systems.

Tanning is a complex process of converting raw skin or hide into refined commercial material 'leather'. It involves treating the skin by applying chemicals, as well as several physical and mechanical processes. Preservation of skin by tanning, and performance of various stages of pre/treatment result in the final product, which has specific practical properties such as stability, appearance, water resistance, temperature resistance, and elasticity. Tanning as a profession is considered to be among the oldest in the world. Leather was also used historically as a means of trading goods.[36] However, its association with caste and Dalits is specific to India. Tanning technology has been developed worldwide, but its caste element has been conserved in India. It has acquired high economic importance, but the work related with tanning is considered low. Caste has coloured the understanding of tanning in India. 'Scientific' and technological explanations have been offered to justify the prevalent caste character of tanning, in which lower castes, their 'lowly' skills, and technological systems have to be necessarily invested. It is argued:

> By nature, tanners are very conservative. This is not simply obstinacy against change; it is because the quality and character of leather is prone to change when the parameters of processing are altered. Changes in the length of processes, process temperatures, float volumes, uptake of chemicals etc. influence the ultimate character of the leather. Leather being produced from a complex, non-uniform natural protein material still requires considerable craft in its manufacture. The adoption of low waste technology often requires a radical alteration of most tannery processes while, at the same time, ensuring that the ultimate product retains its marketable properties. Therefore if a tanner is producing consistent quality of leather which satisfies his customers using a process which may be wasteful in water, energy and chemical utilization, he may resist altering his operations to comply with environmental demands.[37]

Tanning in Jajmau factories has at least four stages: those that are done in the beam house (or pre-tanning operations), in the tan yard, post-tanning, and finishing operations. There is both traditional (vegetable) and chemical (chrome) tanning in the Jajmau area. The traditional tanning takes place in two ways – bag tanning and pit tanning. In bag tanning, the carcass is sewn together into a bag and then tanned with *amla* or *babool* or *myrabulan* bark. The bag is filled with tannin solution and then hanged for many days. In pit tanning, the open hide is soaked in pits and tanned with the vegetable substance. In the pre-tanning tasks, salting, liming, and deliming of the hide or skin are done inside the pits, with traditional or chemical solutions. Traditional vegetable tanning requires hard-core manual labour and also takes a longer time – three–four weeks – for the hide to be tanned. Chrome tanning is considered 'modern', facilitated by powdered chrome as the tanning substance. There are two main stages of the tanning process – production of semi-finished leather through wet blue tanning and crust formation, and leather finishing. There are several sub-stages, like the 'wet-blue' stage when the leather acquires a light blue hue, tanning in chrome liquor, and crust stage as semi-finished leather. Several operations in chrome tanning are done manually. One can find all these operations under one roof, making the working area congested and a curious combination of manual, mechanized, and highly mechanized systems. The present technological system, split into several processes, presents a scenario of traditional and modern, manual and machine, big and small, formal and informal. The tanning process can be split up into many component processes and can be done under a wide variety of production organization forms, depending on how many processes are being undertaken by an enterprise, how mechanized the operations at each stage are, and how employment-intensive and skill-intensive they are, with all these determining how large or small the enterprise is, what kind of employment takes place, and what the conditions of production are.[38]

I observed pre- and post-tanning and finishing products, and talked extensively with tannery labourers in the Jajmau area. Labourers identified key manual works:

a. *Trimming and Sorting*: necessary skin or hide of the animal to be removed from the unnecessary parts, like tail, head, legs, and skin to be classified into different grades;

b. *Dehydration/Storage*: removal of moisture from the skin or hide by applying large salt quantities, as well as cooling the skin;

c. *Soaking*: skin or hide to be removed of salt and other wastes like blood, dung, and proteins by huge amounts of water;

d. *Liming*: removal of hair, flesh, and fat by using chemicals;

e. *Fleshing*: removal of fatty subcutaneous tissues from the skin or hide;

f. *De-liming*: using chemicals to make skin or hide more smooth and adjustable;

g. *Pickling*: applying chemicals for the further tanning stage and for prevention of skin swelling;

   h. *Tanning*: crucial stage through traditional or chemical materials to impart special end-use properties and to add solidity and body to leather;
   i. *Drainage*: wet blue leather to be removed from the excess moisture, mainly by mechanical and manual methods;
   j. *Trimming*: produced skin or hide to be thickened, shaved, and levelled;
   k. *Neutralization*: new bath to the skin or hide in alkaline materials;
   l. *Dyeing*: colouring the skin or hide by pigments;
   m. *Fat-liquoring*: lubricating the skin or hide with an oil coating from animal, vegetable, or synthetic origin; and
   n. *Finishing*: final grooming in specific types.

In Jajmau, there are close interactions of at least three elements of the technological system: the subject, the object, and the tools of technology. The subject – Chamar labour – is everywhere, in its full physical and mental capacity. Labourers told me that their physical labour and skill has immeasurably extended in terms of long hours, and application of mind and body, but the work system has changed comparatively little throughout these decades. The second element – the object – is widespread in the area. Raw materials and the finished products transformed by labour acquire new characters as products of consumption. Touching upon the third element – tools and techniques – I saw that in spite of an increase in area and material, there had not been any substantial change in tools and techniques. It is actually due to the tools of labour, involving hands, forelimbs, and upper body, that the industry has been able to increase its strength many times over. It is not the leather products made, but how they are made and by what instruments that is prominent in the system.

These systems of technology and processes of labour have casteized work in industrial spheres. Technology here has made progress only through particular manual applications of people, harnessing their physical labour within the factory, which is facilitated through a caste economy. Technology and caste function together, where a Dalit has to be occupationally Dalit. As Kanpur tanneries were established industrially and transitioned into commercial forms of production and consumption of leather, the colonial class, army, and emergent industrialists organized factories more and more along social lines. Tannery owners were mostly non-Dalits whereas the labourers were Dalits. Caste supremacy and choice led to a wide adoption of physical labour-based technological systems. Tannery technologies evolved along caste lines, drawing particular castes into certain kinds of work. Even big factory operations were configured on the basis of the labouring body. Factory design, labour tools, training, and education were developed along lines that linked work to caste and body. The story of the leather industry – explained consistently in terms of industry, export, market – became synonymous with caste, culture, tradition, and craft. The entire imagination of the leather industry – its economy, politics, research, communication, and social movement – was coloured by caste. Social ideology and social processes were crucial in the formation of categories of caste and labour, and their interrelationship with state and capital.

Jajmau tanners reiterate that the process of tanning posited the labouring body as a critical site of technological operations. Labourers' narratives of technology had contradictory strands. Some Dalit workers stated that the tannery was their secure source of livelihood in the city, and migration from the village had offered them better work opportunities. Some wanted an improvement in the infrastructure – electricity, water, and housing – so that there was regular work without interruptions. Some felt extremely exhausted and restless in terms of their body, health, and economic situation. Yet, they all carried scars of technology on their bodies. The next section focusses on the condition of Dalit labour in Kanpur tanneries, who have been subordinated to the 'grander' purpose of the leather industry.

Dalit Tanners and Technological Injustices

In his reading of 'the Dalit body', Dipesh Chakrabarty talks of its marginalization in the oppressive Brahmanical schema because of its forced contact with death and waste matter and also because of the relations of oppression that upper castes have built around it being intrinsically connected to the non-human and the non-living.[39] Caste-based social relations reproduce inequalities by devaluing labour of the Dalit body as 'dirty work'. Dalit subjectivities, labour, and sufferings, including occupational hazards, become invisible and 'ungrievable', forcing Dalits to provide a counter-narrative to preserve the memory of their trauma and dignity injuries.[40] In the context of the urban labour market, social exclusion is not just a residue of the past clinging to the margins of the Indian economy. On the contrary, caste favouritism and social exclusion of Dalits and Muslims has infested private enterprises even in the most dynamic modern sector of the Indian economy.[41] The central points of technological injustices – poor people are not regarded as priorities for public investment, their rights to access technologies and technical knowledge are overlooked, their potential as technology innovators is ignored, and technologies are often unaffordable for the poor – have direct implications for Dalit well-being. The imperatives of technology justice demand an urgent paradigm shift in our approach to innovation, technology, and development. It must increase access to technologies and establish new governance mechanisms, which can more effectively curb the use of technologies that adversely affect labour and environment.[42]

The Dalit labouring body remains 'untouchable' in the Hindu caste system – deprived of touch of other humans. However, technological systems of our society became a conduit to 'touch' Dalits and to exploit their body systems for production and consumption. Technology touched Dalits, and paraphrased acts of physical domination and subordination in the everyday life of a factory. Technology worked through a series of sites, where ideas about the body, and how bodies can be extracted were reproduced through the lens of a Brahmanical Hindu society, undergirding operations of caste segregation. Self, body, health, and safety of tannery Dalit labourers inside the factory became centres of my investigation into the evidentiary tropes of technology. Technology becomes visible not only as a tool and technique but also as a category to identify development and destruction of labourers' bodies.

In my fieldwork in Jajmau, I visited twelve tanneries, mostly small and medium, who were conducting tanning processes either in chrome or vegetables, or both. I met and interviewed thirty Dalit labourers who were involved in different operations like trimming, liming, dyeing, and finishing. I did not interview the labourers who were not engaged directly in tanning, for example, office person, guard, and loader. I did not carry any questionnaire and discussed with the workers individually or in small groups of two and three about their machines, work methods, specific tools, personal experiences, health, hospitals, and medical facilities. I did not conduct any physical or medical examination of them. Labourers – all male from Kanpur district and some parts of eastern UP – were of mixed ages, from 25 to 55, and had familial and hereditary backgrounds of leather work. The labourers were mostly tied with one operation. In none of the factories I saw labourers wearing any kind of protective devices like hats, gloves, eye protection, face shields, and foot guards, nor were there any safety devices like machine set-ups, safety relays, and cover walls in and around the machines. Even basic minimums like washing area, water, and soaps were not available. According to the labourers, such equipment and protection was never considered useful for them, and they were simply asked to work with bare hands and feet.

Labourers' bodies – eyes, skin, breathing, stability, and movement – had several visible deformations. Eye redness, pain, irritation and burning were frequently mentioned by the people working in the chrome tanning, buffing, and grinding sections, who were constantly exposed to smoke, dust, and chemicals in the work area. Open skin contact with chemicals in tanning, liming, and dyeing caused burns, irritation, and inflammation, and damaged the skin's natural visual and barrier qualities. There were bruises and marks on hands, wrists, forearms, legs, face, neck, and ears. Skin swelling, redness, and blisters (dry and scaly) were present. They had laboured breathing, coughing, wheezing, dizziness, and weakness. Some also complained of sudden collapse, convulsions, and possibly even paralysis. I met many labourers suffering from respiratory problems. Dust and chemicals in spray painting and chrome tanning sections were breathed in through the air, entering the nose, body, and brain. They had chest pain and symptoms of asthma, with trouble in working, walking, and sleeping. Labourers' bodies, their balance, and movement looked under tremendous strain. In medical terms, problems of musculoskeletal systems – back ache, fatigue, burning, aching, and stiffness – were often referred to because of long working hours in abnormal postures and carrying of heavy weights. In public and visual spaces, they looked crushed under burdens of coercion, patronage, obligation, and indignity.

There have been medical studies on occupational health issues of tannery labourers in the Jajmau tanneries. A study of 497 workers in twenty-one tanneries found that a high proportion of workers (28.17 per cent) had occupational morbidity of four major systems – skin, respiratory tract, musculoskeletal system, and eyes. Non-occupational morbidity was significantly high, with 70.4 per cent workers and 73.7 per cent other tannery staff being affected by several illness.[43] This level of non-occupational morbidity itself speaks of caste of labour, and also includes age, sex, social class, diet, exposures in leisure time, heredity, hygiene, stress, past or predisposing illness and injury, and climate and air pollution. In a cross-sectional survey of health complaints (respiratory disorders, skin complaints, and low-back problems) among 418 laborers in fifteen Jajmau tanneries, it was found that many suffered from low-back pain (61 per cent), asthma (38 per cent), dermatitis (23 per cent), and chronic bronchitis (14 per cent). In general, beamhouse workers reported the highest cases, and chronic low-back pain was significantly elevated, in comparison with workers in the finishing departments. About 44 per cent of the labourers reported at least one period of sickness absence and 17 per cent were involved in a serious occupational accident that required a visit to the local physician.[44]

Another study stated chromium as a major health risk among Jajmau tanners. A high morbidity of about 40 per cent among labourers, compared to the usual 20 per cent, was correlative with high exposure to chromium and its incidence in the blood stream. Exposure to chromium was also behind the higher morbidity of people living in the area, which found respiratory illnesses to be positively correlated with exposure to chromium. Simultaneously, some studies have explored various strategies and methods to promote occupational health in Jajmau,[45] and in various tanning procedures.[46]

These studies have highlighted occupational health and safety issues in tanneries, and ways of addressing them, with a linear health agenda. They have often missed out the physical and social role of the entire technological system in the unmaking of labourers' health and well-being. The mode and application of technology, from its inception to present times, has been critical in the degeneration of labourers' health. Progressive developments in tannery technology from the point of view of hazardous agents like chemical, physical, biological, ergonomics, and safety have not taken care of different stages of hazards management, including identifying, assessing, and controlling risks to labouring bodies. In terms of the interrelationship between caste and technology, divisions between occupational and non-occupational, working and living environment, industrial and social life appear non-existent in Jajmau tanneries.

Many tanners recounted vividly how their low-caste status continued to govern technological systems, and even basic technological reforms have been overlooked, in spite of repeated requests. For example, measures like machines with automatic guards, exhaust systems, water drainage system, and first aid boxes have not been implemented. Grinding operations or tanning processes have not been changed to avoid dusts and solid wastes. The workers referenced other alternative technologies in tanneries – safe, clean, and sustainable, with new skills, diversity, and inclusion – that were now practised in other parts of the world. Labourers severely complained of a majority of tannery works being conducted through piece-rate payment, or prevalence of contract workers in cleaning pits or sludge tanks or industrial sewage lines. Amidst livelihood and wage insecurity, they were forced to work under any system.

The design of the technological system in Jajmau tanneries has been historically conceived exclusively for Dalits – the 'dirty' labourers – and this exclusion gets encoded in the entire imagination and narrative of the sector. Technology has touched Dalit bodies and their everyday lives not to liberate them but to further tie them within their caste identity. For daily employment and survival, Dalit labourers have been deprived of their life and dignity. Despite the exclusive nature of the technological system, and its prominent role in cementing and furthering casteism, and in spite of possibilities of other technologies showing a path to achieve justice, the understanding about tanneries remains limited to occupational health and safety. Dalits have enough individual and collective knowledge about the tannery systems. They have participated in pursuits for improving occupational health and safety within the factory, and they would like to see new systems in place, which are not encoded in old technologies and social hierarchies. However, there is yet to emerge a specifically caste and Dalit approach to redesigning tanneries and ensuring technological justice.

Leather production has been an important activity of society since long. Nature provides little leather to humans in a ready form. The evolution of labour in leather has been determined by material production and the social system in India. Leather production in Kanpur has served various purposes – making useful products for the army and national–international markets. It has also reproduced diverse forms of interactions between labour and resources. Tannery, the core of the leather industry, is a site of labour and technology; it is also a site of social relations, a caste form of production, conditioned by hierarchy and segregation. A casteist economy determines the nature of industry and technology, and civil society culture, of leather. The tanneries are predominantly portrayed as Dalit, in caste and in occupation. The technological and working features of Jajmau tanneries – work, labour, tools, environment, waste, and management – are inseparable from the Dalit labouring body.

Jajmau tanners have been mostly seen from the lens of occupational health and safety, which also speaks of their social positioning. However, in my interviews, Dalits thought highly of their knowledge, skill, and mastery in tanning, but they considered lowly the stigma and casteism attached to their occupation. They explained that their development – employment, income, health, and training – was conditioned by caste, even when the sector was growing well. The caste mode of production constantly reproduces conditions of its existence by torching and deforming Dalit bodies and lives in tanneries. In spite of many changes in terms of capital and ownership structure, and some changes in the composition of labourers, the caste character of Kanpur tanneries is not a transient but a permanent feature. Caste technology, based on Dalit labour, continues to thrive, confining the labourers to captive work systems.

Science and technology have had a complex journey in India, passing through lanes of caste, gender, and nationalism.[47] In the context of post-colonial development, their importance in addressing underdevelopment, poverty, and inequality has been amply emphasized.[48] The introduction of new technology in agriculture and rural development through the Green Revolution in the 1960s has been critically analysed from the perspective of its adverse impact on labour, environment, and health.[49] In the vision of modern India, technology is considered crucial not just for economic development but also for nationalist politics. During the freedom movement, the prominent Hindu nationalist leader Pandit Madan Mohan Malaviya, who shared the political platform of the Hindu Mahasabha and the Indian National Congress, intertwined his views on industry, science, and technology with contemporary political developments, as well as with his efforts to spearhead a Hindu economic and social renaissance. It has thus been stated:

> Madan Mohan Malaviya melded faith and technology to achieve a number of objectives: stimulate fellow Hindus in the capitalist class to realize the ideal of Swadeshi and their dharma of attaining prosperity; shepherd them into a political force capable of articulating the goals of independence, and where required, checking the rise of the Muslim League and the Indian National Congress; and ensure that the social order forged by new technologies did not stray from the religious roots that previously glued communities together.[50]

Technology characterized by caste, too, became a carrier to buttress the Hindu social order. Technology did not merely exploit labour and environment, resulting in various occupational hazards; it was consistently used, altered, and locally revitalized in micro-low-caste communities. Levels of technological application in Jajmau tanneries may have varied, revealing an asymmetric development in small and medium factories, but the basic Hindu caste character of technology remained intact everywhere. Technological systems changed along linear caste trajectories, leading both Dalits and non-Dalits to enter into a vortex of industrial development. Technology has often been on trial for having exploited nature and human beings. However, Jajmau tanneries and its labourers reveal a distinctive character of technology as a multiple and continuous process of producing caste distinction, with a specific relationship to Dalit bodies.

Amongst Jajmau tanneries and Dalit tanners, governmental and non-governmental entities have initiated a range of activities around awareness and information, training and education, research and development, primary healthcare, institutional and legal support, environmental pollution, and infrastructure improvement. However, in spite of their outreach and some positive outputs, they have neglected the profound entanglements between caste, technology, and Dalits in the tannery technological system. The partial 'humanization' of tanneries may have some benefits in the short term. However, they place work within the same caste hierarchy, and in a way that negates ideologies of technological and environmental justice. The new better machines, machine guards, and protective gears mostly strengthened power hierarchies that do not liberate Dalit labourers from their lower positions. As emphasized by Dalit labourers in various interviews, the only way to achieve technological and environmental justice is by moving towards de-Dalitizing and de-casting the tannery technology system.

The upcoming chapter will look into climate change from the perspective of caste and Dalits. It is widely recognized that the destruction of our planet impacts everyone. However, it is important to acknowledge that caste, similar to race, ethnicity, and gender, perpetuates the exploitation, abuse, and even loss of lives within the context of environmental and climate injustice. Certain regions in the country, which have become perilous due to environmental degradation, have effectively become caste-based and Dalit-inhabited zones. Therefore, Dalit organizations are advocating for a distinct caste and Dalit perspective on climate change. This perspective acknowledges the unique experiences, vulnerabilities, and injustices faced by Dalit communities. By highlighting the specific challenges and demands of these marginalized groups, it seeks to address the intersectionality of caste and climate issues.

Notes

1. Gwen Ottinger and Benjamin R. Cohen, eds., *Technoscience and Environmental Justice* (Cambridge, Massachusetts: MIT Press, 2011), 17.
2. Donal MacKenzie and Judy Wajcman, *The Social Shaping of Technology*, 2nd ed. (London: McGraw Hill Education, 1999); Martin Heidegger, *The Question Concerning Technology and Other Essays* (New York; London: Garland Publishing, 1977); Bryan Pfaffenberger, 'Social Anthropology of Technology', *Annual Review of Anthropology* 21, no. 1 (1992): 491–516.
3. E. F. Schumacher, *Good Work* (London: Abacus, 1973), 40.
4. David Arnold, *Everyday Technology: Machines and the Making of India's Modernity* (Chicago: The University of Chicago Press, 2013); Claude Alvares, *Decolonising History: Technology and Culture in India, China and the West 1492 to the Present Day* (Goa: The Other India Press, 1991);

Charles Edquist, *Capitalism, Socialism and Technology* (London: Zed Books, 1985).

5. E. F. Schumacher, *Small Is Beautiful: A Study of Economics as if People Mattered* (London: Blond and Briggs, 1973); David Dickson, *Alternative Technology and the Politics of Technical Change* (London: Fontana, 1975); Godfrey Boyle, *Community Technology* (Milton Keynes: Open University Press, 1978); A. Smith, 'The Alternative Technology Movement: An Analysis of its Framing and Negotiation of Technology Development', *Human Ecology Review* 12, no. 2 (2005): 106–19.

6. Swasti Mitter and Sheila Rowbotham, *Women Encounter Technology: Changing Patterns of Employment in the Third World* (London; New York: Routledge, 1995); Sandra Harding, *Whose Science? Whose Knowledge? Thinking from Women's Lives* (New York: Cornell University Press, 1996).

7. Wajeman Judy, *Feminism Confronts Technology* (London: Polity Press, 1991); Maralee Mayberry, Banu Subramaniam, and Lisa H. Weasel, eds., *Feminist Science Studies: A New Generation* (London; New York: Routledge, 2001).

8. Adam J. Banks, *Race, Rhetoric, and Technology: Searching for Higher Ground* (New Jersey: LEA and NCTE, 2006).

9. Bruce Sinclair, *Technology and the African-American Experience: Needs and Opportunities for Study* (Cambridge, Massachusetts: The MIT Press, 2004).

10. Harold A. Gould, 'The Adaptive Functions of Caste in Contemporary Indian Society', *Asian Survey* 3, no. 9 (1963): 427–38.

11. Raksha Kumar, 'India's Tech Sector Has a Caste Problem', *Rest of World: Reporting Global Tech Stories*, 19 January 2022, https://restofworld.org/2022/tech-india-caste-divides/, accessed on 12 September 2022.

12. Priyanka Pandey and Sandeep Pandey, 'Survey at an IIT Campus Shows How Caste Affects Students' Perceptions', *Economic and Political Weekly* 53, no. 9 (March 2018); Marilyn Fernandez, *The New Frontier: Merit vs Caste in Indian IT Sector* (Delhi: Oxford University Press, 2018).

13. Kanthi Swaroop, 'Manual Scavenging and Technology: Notes on Indian Urbanism', Seminar Series, Indian Institute of Technology, Gandhinagar, 25 April 2022.

14. Interview with Bezwada Wilson, National Convener, Safai Karmachari Andolan, Delhi, 28 June 2021.

15. Anant Kamath, '"Untouchable" Cellphones? Old Caste Exclusions and New Digital Divides in Peri-Urban Bangalore', *Critical Asian Studies* 50, no. 3 (2018): 375–94.

16. Ishika Gupta, Prakashan Chellattan Veettil, and Stijn Speelman, 'Caste, Social Networks and Variety Adoption', *Journal of South Asian Development* 15, no. 2 (2020): 155–83; David Mosse, 'The Modernity of Caste and the Market Economy', *Modern Asian Studies* 54, no. 4 (2020): 1225–71.

17. Esha Shah, *Social Designs: Tank Irrigation Technology and Agrarian Transformation in Karnataka, South India* (Delhi: Orient Longman, 2003).

18. David Mosse, *The Rule of Water: Statecraft, Ecology, and Collective Action in South India* (Delhi: Oxford University Press, 2003).

19. Valerian Rodrigues, ed., *The Essential Writings of B. R. Ambedkar* (Delhi: Oxford University Press, 2002), 159.

20. Mukul Sharma, *Caste and Nature: Dalits and Indian Environmental Politics* (Delhi: Oxford University Press, 2017).

21. V. Geetha and S. V. Rajadurai, *Towards a Non-Brahmin Millennium: From Iyothee Thass to Periyar* (Kolkata: Samya, 2008), 419.

22. Kancha Ilaiah, *Post-Hindu India: A Discourse on Dalit-Bahujan, Socio-Spiritual and Scientific Revolution* (Delhi: Sage, 2009).

23. K. T. Rammohan, *Tales of Rice: Kuttanad, Southwest India* (Thiruvananthapuram: Centre for Development Studies, 2006).

24. Rudrangshu Mukherjee, *Spectre of Violence: The 1857 Kanpur Massacres* (Delhi: Viking, 1998).

25. Syed Asif Ali, 'The Urban Geography of Kanpur' (Thesis submitted for Master of Philosophy, University of London, July 1970), 5.

26. Chitra Joshi, *Lost Worlds: Indian Labour and its Forgotten Histories* (Delhi: Permanent Black, 2003), 62.

27. Charu Gupta, *Sexuality, Obscenity, Community: Women, Muslims, and the Hindu Public in Colonial India* (Delhi: Permanent Black, 2001).

28. Nandini Gooptu, *The Politics of the Urban Poor in Early Twentieth-Century India* (Cambridge: Cambridge University Press, 2001).

29. Ramnarayan S. Rawat, *Reconsidering Untouchability: Chamars and Dalit History in North India* (Ranikhet: Permanent Black, 2012), 115–16.

30. Shivani Kapoor, 'The Smell of Caste: Leatherwork and Scientific Knowledge in Colonial India', *South Asia* 44, no. 5 (2021): 983–99.

31. Shahana Bhattacharya, 'Transforming Skin, Changing Caste: Technical Education in Leather Production in India, 1900–1950', *Indian Economic and Social History Review* 55, no. 3 (2018): 307–43.

32. Rosanne Hoefe, *Do Leather Workers Matter? Violating Labour Rights and Environmental Norms in India's Leather Production* (Netherlands: Indian Committee of Netherlands, 2017).

33. Sumangala Damodaran and Pallavi Mansingh, *Leather Industry in India* (Delhi: Centre for Education and Communication, 2008).

34. Mahmood Alam, 'Problems and Prospects of Leather Industry in U.P.' (thesis submitted for the Master of Philosophy, Aligarh Muslim University, Aligarh, India, 1991).

35. D. D. Kosambi, *The Culture and Civilisation of Ancient India in Historical Outline* (Delhi: Vikas Publishing House, 1975).

36. Subramanian Senthilkannan Muthu, ed., *Leather and Footwear Sustainability: Manufacturing, Supply Chain, and Product Level Issues* (Singapore: Springer, 2020).

37. Shivam Gupta, Rocky Gupta, and Ronak Tamra, *Challenges Faced by Leather Industry in Kanpur* (Kanpur: Indian Institute of Technology, 2007), 3.

38. Damodaran and Mansingh, *Leather Industry in India*.

39. Dipesh Chakrabarty, 'The Dalit Body: A Reading for the Anthropocene', in *The Empire of Disgust: Prejudice, Discrimination, and Policy in India and the US*, ed. Zoya Hasan, Aziz Z. Huq, Martha C. Nussbaum, and Vidhu Verma (Delhi: Oxford University Press, 2018), 1–18.

40. Ramaswami Mahalingam, Srinath Jagannathan, and Patturaja Selvaraj, 'Decaticization, Dignity, and "Dirty Work" at the Intersections of Caste, Memory, and Disaster', *Business Ethics Quarterly* 29, no. 2 (April 2019): 213–39.

41. Sukhadeo Thorat, Paul Attewell, and Firdaus Fatima Rizvi, *Urban Labour Market Discrimination*, vol. 3, no. 1 (Delhi: Indian Institute of Dalit Studies, 2009).

42. Practical Action, *Technology Action: A Call to Action* (Warwickshire, UK: Practical Action, 2016).

43. Abhay Shukla, Satish Kumar, and F. G. Ory, 'Occupational Health and the Environment in an Urban Slum in India', *Social Science Medicine* 33, no. 5 (1991): 597–603.

44. F. G. Ory, F. U. Rahman, V. Katagade, A. Shukla, and A. Burdorf, 'Respiratory Disorders, Skin Complaints and Low-Back Problems among Tannery Workers in Kanpur, India', in *Strategies and Methods to Promote Occupational Health in Low-Income Countries: Industrial Counselling in Tanneries in India*, ed. Ferene Gyula Ory (Les Pailles, Mauritius: Preci-ex Limited, 1997).

45. Ferene Gyula Ory, ed., *Strategies and Methods to Promote Occupational Health: Industrial Counselling in Tanneries in India* (Les Pailles, Mauritius: Preci-ex Limited, 1997).

46. Evgenios Kokkinos and Anastasios I. Zouboulis, 'The Chromium Recovery and Reuse from Tanneries: A Case Study According to the Principles of

Circular Economy', in *Leather and Footwear Sustainability*, ed. Muthu, 123–58.

47. Abha Sur, *Dispersed Radiance: Caste, Gender, and Modern Science in India* (Delhi: Navayana, 2011).

48. Abdur Rahman, ed., *Science and Technology in Indian Culture: A Historical Perspective* (Delhi: NISTDS, 1984).

49. Kartik C. Roy and Cal Clark, eds., *Technological Change and Rural Development in Poor Countries: Neglected Issues* (Calcutta: Oxford University Press, 1994).

50. Arun Mohan Sukumar, *Midnight's Machines: A Political History of Technology in India* (Haryana: Penguin, 2019), xxiv.

8

Climate Change

Weather, Climate, Caste Economy, and Dalit Experiences

At the United Nations Conference on Climate Change at Copenhagen in 2009 (COP 15), Narsamma Masanagari, Manjula Tammali, and Sammamma Begari – Dalit women from Medak district of Andhra Pradesh – burnt their accreditation badges in protest against the lack of recognition of caste and Dalits in climate discussions. Coming from Medak district, where extreme heat and erratic rainfall impacts the everyday village life, Sammamma, a small farmer of Bidakanne village, explained: 'Climate change does make it more difficult. If there is drought or unseasonal rainfall, the first thing that suffers is crop cultivation. If there are no crops, it is difficult for us.' Narsamma, farmer and activist of Pastapur village, claimed that 'upper-caste farmers use machines to plough their land, heightening the climate crisis with fertilizer and other things. Our impact on the climate is much smaller. Larger farmers grow money, we grow food'. Standing outside the Conference Centre, Dalit women demanded 'to bring in the voices of the small and the excluded. If you really want to understand climate change, then come and talk to people like us'.[1] During COP, they also spoke about untouchability and occupational hierarchy still being practised in the villages, and how a group, Deccan Development Society, was trying to organize Dalit women farmers' collectives by pooling financial and human resources, sowing local rice varieties and diversifying crops in small farms, to mitigate the impact of volatile weather and increasing water scarcity. On another occasion, 'Dalits Dignity March' in Delhi, organized every year on 5 December by the National Confederation of Dalit Organisations (NACDOR), a network of grassroot Dalit groups, displayed prominently the banner of 'Dalit Climate Justice'. Ashok Bharti, the chairman of the organization, submitted a memorandum to the prime minister, asking the Indian government not to compromise the interests of Dalits in the country's commitment to cut carbon emissions, and suggested fair share and special care for the community in climate change mitigation policies and budget allocation.[2] Of late, a few Dalit and anti-caste writers have begun articulating their views on climate change. They have critiqued the trend to overlook the 'correlation between climate change and social exclusion', the 'hidden casteism of climate change reporting', and have explored 'a broader framework to situate caste in the climate change/climate justice discourse'.[3]

How does caste matter, and what are the challenges to caste analysis in climate change? Why has there been less focus on understanding caste relations? How do Dalits experience weather and climate? Why does generality about causes and impacts of climate change, even when formulated in terms of the poor, the marginal, and the vulnerable, not capture sufficiently caste and Dalit-specific situations? How does a framework of Dalit 'vulnerability' and 'protection' on the one hand, and glorification of their traditional caste-based work as 'smaller carbon footprints' and 'mitigation of greenhouse gases' on the other, reinforce the dominance of powerful castes and classes in climate politics? What are the definitions and meanings of climate justice for Dalits and low-caste people? How do Dalit organizations ground their politics of climate change, which differs from the mainstream environmental and civil society groups?

Such questions should be raised, and a Dalit lens should be taken to discuss some aspects, which have been marginalized in climate change debates. I particularly focus on the intersections between caste, Dalits, environmental justice, and climate change, which include everyday experiences, history, economy, labour, employment, and health on the one hand, and an absence of low-caste representation in science, technology, policy, and governmental setting on the other. Dalit life narratives are preoccupied with weather and climate as central to their human condition. The caste economy is intimately interwoven with climate-related natural and health hazards in degraded occupations and dangerous locations. Environmental organizations, climate science, and technology are in denial of caste structures, which have resulted in the imposition of unfair and unequal burden on Dalits. Climate justice is inextricably linked with caste injustices.

However, this is not to deny that climate change can have a universal impact on society, cutting across caste divisions. Caste relations, and the position of Dalits within it in particular, signify socio-cultural constructions, occupations, roles, and responsibilities. Caste is diverse and fluid, with significant changes over time, space, and region. However, its persistence and endurance has deep religious and social roots. For example, spatial divisions between Dalits and *savarna*s have impacted the health and life status of low-caste people. Occupational roles and responsibilities, education, opportunities, autonomy, and mobility among different castes play a part in determining their vulnerability to effects of climate change. Caste divisions in labour and work around the country have important bearings on our knowledge about the common but differential impact of climate change between and within communities.

Climate Change in India: Perspectives and Predicaments

Climate change has been extensively researched by academics and activists, government and civil society organizations, corporate and business bodies, policy makers and politicians. Its literature covers a diverse range of subjects and issues, including the science of climate change and the nature of its possible impact on people, regions, and sectors; climate colonialism and the differentiation between north and south; bearings of capitalism, industrialization, and globalization on it; contests and compromises, continuities, and changes in climate negotiations, and a critical analysis of India's official positions; debates on environment, development, equity, and justice; challenges of mitigation and adaptation; comprehensive views on sectors like power, energy, transport, construction, urban development, agriculture, water, and forests; domestic politics, and the role of political parties, civil society organizations and campaign groups; Indian business, industry, Clean Development Mechanism, technology, and labour; and much more.[4] With increasing concerns on climate change, the diversity and depth of research on science, development, vulnerability, agreements, fossil fuels, finance, technology, renewable energy, policy integration, and future agendas has further developed.[5]

Multifaceted issues of poverty; poor, marginal, and vulnerable communities; inequalities and injustices; gender and power relations; domination and dignity; human security; human rights; and freedom have also been discussed to understand the past, present, and future of climate change at international, national, and local levels.[6] The National Action Plan on Climate Change (NAPCC), released in 2008 by the Government of India, had its first 'principle' as 'protecting the poor and vulnerable sections of society through an inclusive and sustainable development strategy, sensitive to climate change'.[7]

At the United Nations Summit in 2015, the Indian prime minister Narendra Modi stated: 'When we speak of climate justice, we demonstrate our sensitivity and resolve to secure the future of the poor from the perils of natural disasters.'[8] Simultaneously, Indian civil society organizations and networks continued to raise 'concerns about the adverse impact of global climate change on the poor and most vulnerable section of the society', and declared their commitment to promote 'equity and social justice between peoples, sustainable development of all communities and protection of the global environment'.[9] Thus, the poor are visible and recognized in climate change discussions, with an attention to its impact on inequality, exclusion, vulnerability, rights, and justice.

Social science research in India has amply analysed how caste works in the Indian economy, poverty, environment, village, family, politics, and culture. Logically, therefore, caste and Dalits should also be considered necessary to understand the human, social, and economic dimensions of climate change. However, concerns around caste and Dalits have largely remained invisible and unrecognized in climate change debates.[10] There are several links to establish how India's Dalits are structurally, economically, and socially disproportionally vulnerable to climate change. For example, India's majority population lives in rural areas, where land, agriculture, and related livelihood activities are critical for economic and social security. Land is also associated with caste, power, identity, and mobility of people. Landlessness affects everyday survival, access to governmental schemes, and the capacity to resist poverty and disastrous situations. However, the agricultural landholdings of different caste groups in India are highly unequal. About 79.33 per cent operational agricultural land is owned by 'Others', including dominant castes, while Dalits and tribals, who are 20.6 per cent of the country's population, own only 11.84 per cent. Further, if we consider the differential distribution of landholdings among different castes, the disparity is more pronounced; Dalit landholdings are mostly of small–marginal sizes. Out of the 11.91 per cent of land owned by Dalits, 78.19 per cent is marginal land (that is, below 1 hectare), and less than 1 per cent of Dalits own land of 10 hectares or more. Thus, a majority of landowning Dalits are marginal landholders. In the rural sector, 52.6 per cent Dalits are casual labourers.[11] This unequal land ownership reduces land productivity and sustainability, and in times of flood, drought, natural disasters, and climate change risks, landless Dalits are more vulnerable than other castes.

Similarly, the pathetic rural housing situation for Dalits, in terms of shortage–availability and basic amenities, makes them physically and mentally vulnerable to changing climate situations. The Census of India (2011) and

National Sample Survey Organisation (NSSO) 69th (2012) data found that though 94.7 per cent Dalits have self-owned houses, only 43.8 per cent have permanent houses, 34.5 per cent houses possess a separate kitchen, and 38.4 per cent have safe drinking water available. Thus, the quality of Dalit houses – tenancy of homestead land, condition of house, number of dwelling rooms, availability of amenities – is poor, making them all-weather susceptible and unsustainable.[12] A disjoint between the common cause and impact of climate change on the one hand, and caste cause and impact along with Dalits and low-caste specificities on the other, highlights the skewed social understanding of most of the climate experts and scientists in India. This kind of social generality has three main strands: it is 'caste-neutral'; it believes in certain 'caste stereotypes'; and it implicitly reveals a 'climate casteism'. I will explain these briefly, to show how caste is at work within climate experts, consciously or unconsciously, overtly or covertly.

'Caste neutrality' is a classical political approach, which focuses on human poverty and inequality generally, without considering individual caste identity. Poor people are perceived in broad terms as suffering men and women, who inhabit diverse climate zones – dry or wet, hilly or coastal, fragile or disaster-prone. They are victims – living dangerously and dying, migrating and adapting to the impact of climate change. A simple word search of the NAPCC shows that the words 'caste' and 'Dalit' do not appear even once on its pages – they mostly refer to 'human' (mentioned 19 times) and 'poor' (7), and sometimes to 'women' (4). Apparently, India's Ministry of Science and Technology is interested only in the broad category of human (3) and poor (9), not even in men and women, in 'climate change and agriculture', while discussing the impact, adaptation, initiative, and way forward.[13] Research and publications by well-known climate change experts and civil society organizations are identically similar, except that they do mention 'human' and 'poor' more often in their pages, but again without a single reference to caste or Dalit.[14] These works have been important in anchoring our understanding of climate change in the context of poverty, inequality, and marginality. However, they have a blind spot about the interrelationship between caste, Dalits, and climate change.

'Caste stereotype' is a prevalent climate approach that locates low castes and Dalits amidst their poverty, degraded occupation, illiteracy, and vulnerability, and constructs from there that they have less consciousness and agency. In typical and conventional ways, economic vulnerability is considered as equal to climate vulnerability, and poverty is identified as the most significant socio-economic variable for the low-caste communities. Thus, poverty reduction and economic development are taken to be the main pillars of adaptation and mitigation strategies for 'weaker' and 'low-income communities', by both government and civil society initiatives.[15] Further, present and future climate variability of Dalits is analysed, without taking into account the changing dynamics of employment, occupation, education, skill, mobility, and community or inter-caste relation. Stereotypes of ignorant, illiterate, idle, and unintelligent Dalits and low-caste people also determine the nature of their role in climate change policy formulation and implementation. This approach is biased and partial not only because of its negative socio-cultural stereotypes about the historically disadvantaged castes but also because of an absence of analysis of entanglements between caste, poverty, and climate variability. It is unconcerned about whether, why, and how caste and poverty intersect with climate change policy and politics.

'Climate casteism' involves representing, reworking, and formalizing degrading and dehumanizing subsistence and polluting work, labour, and structure for mitigating climate change. Caste structure makes certain kinds of labour and work segmented and stigmatized, and the untouchable labouring body bears the brunt of physical and social exploitation. Physical labour in leather and waste collection, sanitation, and rag picking, among many others, is idealized or romanticized in this approach. In the process, labour is separated from pain and revulsion, which is actually a part and parcel of Dalit performance. In fact, such caste-insensitive climate sensibility sustains caste hierarchy in various initiatives, policies, and governance practices. For example, NAPCC hails the informal sector of waste picking and recycling as 'the backbone of India's highly effective recycling system', and wishes to strengthen the 'informal sector systems of collection and recycling'.[16] Similarly, several environmental organizations meticulously calculate the mitigation of greenhouse gases by informal waste pickers. They are characterized as 'cooling agents', 'climate entrepreneurs', and 'climate champions', who should be given rights, protection, and incentives to do the same job that the Manu code has determined for them since centuries.[17] It is no surprise that such climatality has no reference to its embodiment in caste and polluted Dalit bodies.[18]

Climate change scholarship in India has been vivid and vibrant, and it is not possible to classify it in neat, exclusive categories. However, caste and Dalit as categories of analysis have been largely missing in most of these studies. Dalits are thus usually merged in the general categories of poor, marginal, or vulnerable. In fact, on some crucial issues of climate change, like weather and climate, adaptation and mitigation, there appears to be no reading of a specifically Dalit experience and narrative. If rarely Dalits do make an appearance, their perspectives are elided. I now deal with Dalit narratives on weather and climate, in order to explore their complex relationship with nature.

Weather and Climate in Dalit Life Narratives

It is your human environment that makes climate.

– Mark Twain[19]

In *The Weave of My Life: A Dalit Woman's Memoirs*,[20] Urmila Pawar narrates the travails of her weather life in the Konkan region of Maharashtra. She travelled early morning to the Ratnagiri market to sell various things, by crossing over rocky land and dangerous hilly slopes of the Sahyadris ranges running along the coastline. It was a harsh landscape – howling winds blew continuously, ferocious enough to topple a person to one's death, and fear of injury often made Urmila's heart cold with terror. In the rains especially, her life hung by a thin thread. Urmila remembers the rains:

> Survival was as dicey as gambling. The rain pouring down in huge torrents, as if the sky itself was collapsing; lightning striking across the sky in deafening roars; streams fiercely gushing down the hills; rocks exposed under receding layers of soil, like teeth jutting out of monstrous mouths, ready to tear the traveler's feet to shreds; thick shrubs, huge trees with wild creepers weaving tangled webs, and dense, dark forests sprawling around; rivers swollen with floods, weathered wooden bridges over them, ready to collapse at any moment.[21]

Dalit writer and activist Vasant Moon experienced dead heat in his *vasti* (neighbourhood) in Nagpur city in the Vidarbha region of Maharashtra.[22] He describes how summer began to burn since March; by April it was impossible to go outside after the morning; and by May the roads were empty in the afternoon. The turbulent air would give impetus to whirlpools, and then the harshly burning winds would lift up the dust in the field and blow it into the neighbourhood. The hot blast of air would subside at night and only after two in the morning, when a clear, tranquil wind provided a cooling touch on their bodies, could they sleep. The summer was always stubborn, but people were equally hard-headed. Moon also details out the joys of the monsoon – thick clouds, first rains, whirling birds, playing in the flowing water, heavy showers, drains overflowing, catching and cooking the crabs, and celebrating the welcome change in weather.

Weather and climate,[23] key markers of climate change in contemporary times, have been central to Dalit life. In a large number of Dalit life narratives,[24] weather patterns and climate systems often appear, and can be taken as lighthouses to enhance our understanding of climate change. Dalit experiences have their own characteristics, which are usually defined differently from dominant climate discourses. Identifying these is not only to sketch the victims and the vulnerable, and the cluster of causes and effects, but also to recognize the ambiguities and complexities around meanings of weather and climate, and their close connections with Dalit life. Dalit life narratives reflect the compound reality of living with weather over a period of time, spread over months and years. They experience climate change through everyday and cataclysmic events. Caste overwhelmingly determines their weather and climate sense.

Day-to-day functioning of the atmosphere, like temperature, rainfall, wind, humidity, sunshine, and cloud, are heavily filtered through living and labouring conditions. A cold morning, a sunny afternoon, a rainy day, or a chilly night can be generalized in broad descriptions. However, their experience and impact are never static and are often caste- and Dalit-specific. Urmila Pawar is a Mahar. In her village, poor Mahar women had no choice but to journey through a difficult road amidst slopes and hills, with heavy bundles on their heads, filled with firewood or grass, rice, or semolina, long pieces of bamboo, and baskets of ripe or raw mangoes. Animals attacking people on their way were not uncommon. Women were also occasionally assaulted by men hiding in shrubs and trees. In Pawar's description, women and weather together acquire a particular state of being:

> With their emaciated bodies covered in rags, bony sticklike legs, bare feet, pale, lifeless faces dripping either with sweat or rain, sunken stomachs, palms thickened with work, and feet with huge crevices like a patch freshly tilled, they looked like cadavers floating in powerful streams, propelled by a force hurtling them along the strong currents, being dashed against rocks and thrust forward by powerful waves.[25]

Various weather elements continued – day by day, year by year – and created a combined weather–climate regime, not in a narrow sense of average weather or statistical description but in a wide set of the complex human–climate system, consisting of interactions between labour, land, atmosphere, exploitation, development, and change. Pawar witnessed the working of such a regime in her village, in her home, and in her everyday life. Her tiled roof house constantly got hotter; the firewood stove in the kitchen was completely blackened with smoke; food had to be gathered from the forest, river, and sea even in extreme times; and cooking and cleaning done amidst water scarcity. Changing climate and its impact was very much visible here and now:

> Gradually the tide would flow in, water rising from all sides. Then they yelled warnings at each other, hurried one another, swiftly cut off huge pieces of rock with oysters in it, filled their baskets, trying to collect as much as they could. In the hurry sometimes novices would simply pick up rocks thinking they were oyster shells. When the water rose to their knees, they would hurry out of the water, balancing the heavy basket on their heads. Some women lost their lives because they didn't notice the water rising. My maternal uncle's daughter Lakshi was saved from drowning by some people. But they could not save the two other women with her who drowned.[26]

When Urmila later visited the village of her husband, the climate was no different – barren lands, parched dry grass and shrubs, and land-owning farmers burning their land as preparation for the next planting season. In Vasant Moon's portrayal of 'neighbourhoods', 'heat and rain', 'falling star', 'callousness and clouds', the unmitigated weather–climate regime has a life and death implication. He views it as a natural, human creation, and a matter of caste and location, emergency, and fortune. Accounts of heat waves, shivering nights, and overflowing rivers abound. He narrates a story of sudden flooding of the river Wainganga, and the courage of a young girl, Nathi, who swims through the fast-flowing waters. She defines an individual's capacity to express freedom from climate circumstances, as she instinctively decides to express her independence by challenging the stream.

Dalit lives show an acute awareness of the complexities of weather, climate, and climate change, where caste distinctions play a critical role in shaping their experiences. In a Dalit life, climate can be the sum total of a dynamic state of living, where temporal and permanent, every day and episodic, variable and stable, extreme and ordinary can co-habit at all times and locations. Jeya Rani explains sharply the intersections of weather, climate, and caste: 'Dalits are not afraid of climate change or any other natural disasters, because socially and economically, they have been leading a disastrous life forever. For them, disaster caused by the upper-caste Hindus is the real problem.'[27] Dalit experiences of weather and climate are further amplified through a case study of the brick kiln industry in Haryana, and a sociological analysis of caste relations in climate change.

The case study is based on my fieldwork in June–July 2018 in six small and medium-sized brick kilns of Passour, Badli, Kablana, Khungai, and Kanonda villages of Jhajjar district in Haryana. These villages are known for their brick kiln industries. I thus selected them for my field work. I have used a qualitative, case-study methodology through field visits, interactions, and observations within the brick kilns and labourers' homes in the area of study. I have travelled five times to the area and interviewed thirty brick kiln labourers (ten women), twelve villagers, and representatives of labour and Dalit organizations. During my fieldwork, quantitative and qualitative evidence and data was collected through extensive interviews and group interactions with labourers and representatives of government departments and civil society organizations in the city. In the villages, I was often accompanied by a local resource person. In different villages and in Jhajjar city, I cross-checked and cross-referenced facts and situations that were mentioned in the interviews. These were then supplemented through other published records and printed sources. Academic writing on brick kilns and labouring population, governmental, and non-governmental reports were also included in the research.

Understanding Climate Change through Caste Relations

Haryana's Jhajjar district is 50 kilometres away from the national capital of Delhi.[28] A network of national highways and railway stations connect the district to neighbouring cities likes Gurgaon, Rohtak, Bhiwani, and Rewari, and to the states of Delhi and Rajasthan. The district is predominantly rural and agricultural in nature. The 2011 Census counts a total of 185,334 households in the district, of which 136,503 are rural, which are largely engaged in cultivation and agricultural labour.[29] Since the early 1990s, districts like Jhajjar, Bahadurgarh, Badli, and Beri also became important industrial sites for large, medium, and small industrial units in diverse sectors like leather goods, sanitary wares, chemicals, plastics, electronics, steel wires, utensils, pharmaceuticals, paper products, and glass wares.[30]

In the 1990s, brick kilns, too, developed as an important industry in Haryana.[31] In particular, Jhajjar became known for its large production of bricks – red, fly ash, clay face, handmade, and perforated. The district has more than 400 brick kilns, the highest in the state, which are owned by local and regional manufacturers, known variously as Raj Singh Bricks, Mehar Bricks, Sri Sai Enterprises, Shiv Enterprises, Hanuman Bhatta Company, Shiv Bhatta Company, Jai Maa Laxmi Builder, and many more. The industry employs around 50,000 local, seasonal, and migrant labourers every year between October and July. The urban and rural landscape of Jhajjar is often identified with smoke-billowing chimneys of brick kilns, tin-roofed sheds stuffed with rows of freshly moulded bricks, and 6-foot-high shanties of workers covered with plastic roofs and loosely arranged brick walls. Traditional brick kilns are energy intensive and a source of greenhouse gas emissions. A study of labourers, subjected to occupational and metrological heat stresses in hazardous situations, unravels a journey of temperature, heat, greenhouse gas emissions, and their specific caste origins, along with Dalit links to this chain. Intersections between socio-economic injustices and climate change risks show how caste matters in heating up temperature, and the placement of Dalits in anthropogenic climate situations.

India is the second largest producer of bricks in the world, manufacturing roughly 240 billion bricks annually. There are around 140,000 brick kilns that contribute more than 10 per cent to the total brick production of the world. They employ about 10 million workers and consume around 25 million tons of coal annually.[32] Brick kilns have also been documented extensively for their severe violations of labour and human rights, including of women, children, and bonded labourers.[33] A majority of the kilns belong to the informal sector, operate in an unregulated manner, and employ migrant labourers. Systems of contractors, bondages, advance payments, loans, and compound interests are widely prevalent here. Research recently analysed the heat risks for migrant workers at brick kilns. It combined occupational conditions and climate change, and suggested local technical solutions and appropriate technologies, which can be applied to mitigate worsening heat stresses.[34] Another report strikingly studies heat risks in brick kilns for labourers and animals together, amidst hazardous and tough situations, and states unambiguously:

> It's hard work for both equines and people who work there. For equines, the harsh environment causes serious health problems such as disease, injuries and lameness, and with temperatures exceeding 120°F, heat stress is also an issue. It's a serious condition which, if left untreated, can lead to life-threatening heat stroke.[35]

Of late, several governmental and non-governmental organizations have initiated field research and various measures in the Indian brick sector, with a perspective of reducing emission of high air pollutants and introducing cleaner raw materials and fuels.[36] At the same time, caste and Dalits are closely aligned to the brick kilns in Haryana.

Jhajjar is overwhelmingly agrarian, where the dominant caste of Jats owns a majority of the agricultural land, including big, marginal, and small holdings. Due to a relative availability and facility of irrigation, water, electricity, road, transport, and market in the region, crop intensity in the district is 152 per cent, and every major kharif and rabi crop is grown here.[37] However, Dalits, who are 17.78 per cent of the district's population, have virtually no agricultural land. With increasing industrial, trade, and business activity in Jhajjar district since the 1990s, Jat farmers have diversified. They use their non-irrigated and faraway agricultural land for brick kilns, in order to create another source of income with minimum expenditure and least risk.[38] Brick manufacturing companies based in Haryana, Punjab, and Delhi, with their networks of local contractors and suppliers, have flourished in these Jat lands, who have either leased their land to these companies or have acquired licences themselves. This has transformed the process of brick making, which was hitherto mostly *maidani bhatti* (awa) – small, scattered field structures used for firing bricks, in which neither a chimney was used nor slack coal consumed as fuel. Local-level manufacturing and consumption of bricks also required lesser amount of ordinary clay, which was even possible to procure sometimes from village ponds and wastelands. For these reasons, until the early 1980s, manufacturing of bricks in kilns of Haryana was covered by the Food and Supplies Department, which was under the Haryana Control of Bricks Supplies Order, 1972.

Topographically and climate wise, Jhajjar forms a part of the Indo-Ganga alluvial plain, and is dotted with sand dunes, small, isolated hills, and flat lands. Hot summer, cold winter, and meagre rainfall are the main climatic characteristics of the district.[39] In such a context, brick kilns, generally of medium and large production capacity (2–10 million bricks every year), were established in clusters around Jhajjar, Bahadurgarh, Badli, and other peri-urban areas. Over time, landscapes of widely spread-out brick kilns, changes in land use/land cover, degraded–dry areas, manual labourers and animals, trucks and tractors, use of coal and some biomass fuels have become familiar sites in the region. The brick production process has remained manual and seasonal. It has also become highly energy-extensive, with huge environmental footprints. Work process is divided into specialized categories, and labourers, too, are divided accordingly. *Paatla*, that is, raw brick making, is the first process, where a brick mixture is made with earth and water and set into a mould to be dried under the sun. This is followed by *bharai*, that is, shifting the sun-dried raw bricks to the firing kiln for firing; *khadkan*, that is, arranging the raw bricks in a specific style in a stacked array for proper firing; *jalai*, that is, firing in the kilns in high temperatures; and *nikasi*, that is, after cooling, removing the fired bricks from the kiln and transporting them to wooden hand-propelled carts, to be finally loaded in trucks for supply. These work processes are labour-intensive, for which high levels of human energy are required. In climate-sensitive areas, energy-sucked labourers have always been more vulnerable to temperature and heat risks.

Brick labourers in Jhajjar are Dalits, who belong mainly to Chamar, Dhanuk, Valmiki, Dagi, Deha, Gagra, Sansi, Khatik, Pasi, Od, and Meghwal scheduled castes, and come from the states of Haryana, Punjab, Bihar and Uttar Pradesh. Men, women, and children work together; some work categories are male-dominated, whereas some others are female-centric. In terms of caste-based occupation, brick kilns of Jhajjar can be compared to the sanitation sector, where only Dalit labourers are found. Though not considered polluted, brick work is seen to be as degraded and lowly as sanitation. While there have been several reports on labour exploitation and human rights violation in the sector, the impact of temperature, heat, emission, and climate on Dalit labourers specifically has been marginalized.[40]

'It is hot as hell'; 'it boils inside'; 'these are dog days'; 'living to death' – these are some of the expressions of Dalit labourers working in the brick kilns of Passour, Badli, Kablana, Khungai, and Kanonda villages in Jhajjar district. Radiant heat from the kilns in which bricks are fired, and overall working conditions, are worst between May and July, when temperatures climb to 45°C. It is a 'fate worse than death', as excessive heat has severe adverse effects on Dalit labouring bodies, and their function, capacity, and capability is impacted in visible and invisible ways. Heat shocks and strokes, sometimes fatal, are regularly reported. Dalits complain of heat rashes, cramps, exhaustion, dizziness, eye and body burns, headaches, fever, and injuries, which are exacerbated due to poor cooling options and high workloads. Our fieldwork suggests that even amidst heat stress, there is no slowdown of work, break, or limit in the number of working hours, also because workers are paid on piece and time-rate basis.[41]

Coal and other biomass fuels in brick kilns result in emissions of particulate matter (PM), including black carbon (BC), sulphur dioxide (SO_2), oxides of nitrogen (NOx), carbon monoxide (CO), and carbon dioxide (CO_2).

These emissions have a severe impact on health, climate, and vegetation. In the recent past, high-ash, high-sulphur coal, industrial waste, and loose biomass fuels have also been increasingly used in brick kilns (due to higher cost and shortage of good quality bituminous coal), thus resulting in new air emission challenges. In our interviews, labourers repeatedly stated regular respiratory problems, often leading to serious illnesses like bronchitis, pneumonia, asthma, and pulmonary disorders, and lack of medical facilities.[42] Labourers specifically mentioned the heat wave season that they are frequently experiencing since the past few years. The 2015 heatwave finds special mention. In the summer of 2015, from late May to early June, north-western, central, and coastal India experienced severe heatwaves, when maximum temperature exceeded 45°C for many days and caused loss of thousands of human lives.[43] According to labourers, women workers suffered exceptional heat stresses in 2015. Heat, sun strokes, and dehydration were very high among all labourers. However, there has been no reporting of specific heatwave-related issues in brick kilns.

A long-term analysis of temperature trends in Haryana shows an increase of 1°C to 1.2°C in mean maximum and minimum temperature. Many parts of the state, including Jhajjar, show a decreasing trend in monsoon rainfall. It is projected that by the 2050s, climate change in Haryana can lead to a mean maximum temperature increase by 1.3°C, and mean minimum temperature by 2.1°C, which will severely impact water resources, forests, agriculture, and livestock.[44] There are clear warnings that an increase in air temperatures and heatwaves, combined with air pollution, will lead to an escalation of deaths from cardiovascular and respiratory diseases in the future. Jhajjar figures prominently in increasing minimum and maximum temperature, variations in climate, decreasing monsoon rainfall, and heatwave projections. However, the underlying caste dynamics of this climate change finds no mention. While traditional production of bricks through small-scale, decentralized kilns, and by using natural resources of the village, dates back to the post-independence history of Haryana, brick kiln as an industry and brick as a big source of money and revenue evolved during the 1980s. Jat landowners became heavily involved in the industry, and the Jat–Bania–Brahmin combine took control of the brick industry of the state. Under them, the quantum of brick production sky-jumped, and the production itself became more extractive, polluting, and hazardous, leading to high greenhouse gas emissions and air pollution. Dalit labourers, who faced social discrimination, labour and human rights violations, violence, and repression, had to also bear the brunt of increasing temperature. Dalits are more exposed to carbon emissions and heat, which are part of the new industrial order. Dalit labour is trapped in caste and climate paradoxes, which underline the reality of labouring within a hierarchical social order. In light of this perspective of caste relations in climate change on the basis of a case study, let's now turn to how perspectives from caste and Dalit studies can address the caste and climate question.

Caste Economy and Climate Justice

The climate and caste question have been understood mainly from the perspective of vulnerabilities, adaptabilities, inequalities, impact, and resilience.[45] While it may be right that Dalit and low-caste people suffer more because of lack of resources and power, the prism of discrimination and resilience does not suffice for caste understandings of climate change. Feminist environmental scholars have critiqued discourses of environmental security that propose masculine, strong, state responses to combat climate change.[46] Analysing the emergence of a 'vulnerability-resilience' dualism that now dominates climate policy at all levels, Sherilyn MacGregor has argued:

> This move reflects the hegemonic masculinist framing of the crisis. Climate victims appear as passive figures in need of help to become self-reliant enough to cope in harsh conditions beyond their comprehension and control. People who are positioned as vulnerable to extreme weather and other forms of climatic destruction are both feminised and racialised.[47]

The interrelationship between race, class, and climate change in the United States has been studied to analyse how black Americans under systemic racism are more vulnerable to natural disasters such as Hurricane Katrina in 2005 and Hurricane Maria in 2017, and lack support in post-disaster rebuilding and recovery.[48] Similar to feminist and black critiques of dominant climate change discourses and policies, a caste perspective can foreground knowledge of the working of caste structures in economy, power, science, and technology in accelerating climate change. This frame of reference can focus on the nature and structure of production of goods and services, and the organization of economies to create specific work profiles along caste lines. Economic scholarship on the interrelationship between economy, caste, and Dalits in post-independence India can certainly throw light on how caste is being restructured to perpetuate discrimination and disparity in economy, with upper castes at the top and scheduled castes and scheduled tribes at the bottom. This scholarship also highlights that the caste-based local economy is oriented to serve the interests of big capital and markets under new global–liberal regimes, which accelerate climate change. Ashwini Deshpande takes a macro, all-India view, and uses large data sets to demonstrate the continuing 'caste–occupation nexus' and 'contours of discrimination in the modern Indian economy'.[49] Based on a detailed field study of caste and business associations in the market town of Arni in northern Tamil Nadu, Barbara Harris-White conceptualizes 'caste-corporatist capitalism' that explains ways in which 'caste is being reworked in the contemporary era to be an instrument of corporate regulation'.[50] Caste works as an economic regulator, and the caste system and its ideology 'form a significant component in the local structures of accumulation'.[51] On the basis of ninety detailed interviews of Dalit entrepreneurs in thirteen districts of six states, Aseem Prakash shows how caste predominantly helps the accumulative endeavours of upper castes in markets. According to this study, 'Caste takes an organised form to connect individuals, articulate their economic and political interests, set norms of market behaviour, regulate credit and labour supply in the market, help in procuring market contracts, set prices, regulate the entry of new market players, etc.'[52]

These studies provide a conceptual framework, as well as ground knowledge, on the development of market economy and interlinked processes of climate change – production, distribution, consumption, competition, and technology – understandings of which have been largely shaped by dominant discourses. Likewise, the climate change discourse reflects the dominant power structure that is casted in caste. The caste economy has also to confront issues of equality, participation, and representation in decision-making processes regarding climate change. There is substantial literature on the evolution of business communities and capitalists along caste and community lines in post-independence India, and their preferences in terms of regions and sectors.[53] In comparison, there is little work on how caste ideas, practices, and values influence environmental concerns in economic arenas. It has been argued elsewhere that male-dominated and women-represented groups have taken different decisions regarding environmental concerns. Data-based studies also suggest that conservative white males and elite white men are significantly more likely to deny climate change than other Americans.[54] The hyper-masculine character of the banking industry is being pinpointed as an important reason for ignoring signs of finance and climate crisis.[55] Similarly, an analysis of the *savarna* effect on economic decision-making pertaining to climate change and environmental issues is relevant to see which values are leading to what outcomes in resource extractions and carbon emissions.

Amidst this background, the contours of climate justice can be revisited to include Dalits specifically. Climate justice has dynamic definitions globally and locally, which are anchored in environmental, social, distributive, and energy justice; intersection of environmental injustice with class, gender, and race; inequality at transnational, intergenerational, and local levels regarding risks and burdens; and reclamation of human dignity, human security, and universal human rights.[56] The bedrock of climate justice politics has been established by focusing on experiences, issues, and visions of a majority of people for a climate-sustainable world. Movements around climate justice have committed themselves to focus on the marginal and the local, who are most impacted by climate change. Phrases and slogans like 'indigenous climate action' and 'no climate justice without gender justice' have become popular. For climate justice, the caste–climate paradigm is as deep and critical as gender–climate, for equality and justice, representation and participation on the one hand, and to unravel exploitation of resources, work, and wealth in the Indian economy on the other. In this context, it is pertinent to describe how a Dalit organization – National Campaign for Dalit Human Rights (NCDHR) and its offspring – has evolved their climate politics, by linking it to the ground reality, community, research, and advocacy. They have tried to connect the 'core' issues of climate change and justice with the experiences of their constituents.

NCDHR and the Grounding of Climate Justice Politics

Founded in 1998, NCDHR is a coalition of Dalit human rights activists and academicians, working to annihilate casteism and caste-based discrimination. With branches in fourteen states, the organization has classified its movement into five fields – gender justice, economic justice, access to justice, disaster risk education, and global Dalit rights. For this it has initiated a set of organizations – All India Dalit Mahila Adhikar Manch, Dalit Arthik Adhikar Andolan, National Dalit Movement for Justice, and National Dalit Watch.[57] Since its inception, the network has wished to make Dalits visible in national and international arenas, and to hold the state accountable for human rights violations committed against Dalits. The Indian Ocean tsunami of 2004 and its aftermath – immense loss of lives, homes, livelihood, environment, and discrimination against Dalits in relief and rehabilitation – led NCDHR to strategize and ground its climate work, by focussing on community mobilisation and entitlements, and research, publication, and policy engagements in the framework of social justice and human rights. NCDHR first began to address the exclusion of Dalit communities during disasters, and simultaneously started drawing its connections with issues of livelihood, land rights, agriculture, and food security.

Paul Divakar, a key organizer of NCDHR, suggests Dalit models for conceptualizing and grounding climate justice, which address the everyday of weather and climate through a variety of interventions. The daily existence of Dalits – their labour, place, occupation, discrimination, violence, and the intensity of living in hard and hostile situations – are the real issues of climate change. The politics of climate justice sometimes is distanced from the everyday realities of Dalit lives. States Divakar:

> Dalit issues are climate issues; climate change is about us. Even before the advent of so-called adaptation and mitigation, Dalits had been negotiating and challenging climate change impact. One way or another, droughts, floods, and natural disasters have been ever-present in Dalit lives. Local mobilization and resistance to caste injustices, analysis of climate change as an issue of human rights, and related advocacy and campaign for policy change should be our focus. However, when we raise Dalit–climate issues, we are declared as 'us' versus 'others'. It has been a slow and difficult journey to find our feet in this field.[58]

The climate work of NCDHR revolves around challenges of inclusive disaster recovery and rehabilitation for Dalits and marginalized communities, making of an inclusive disaster risk reduction (DRR) policy, law, and guidelines that recognize caste discrimination and exclusion, creating a constituency of human rights defenders to protect rights and entitlements of disaster-affected people, and develop tools, methods, and capacity for just DRR actions. The organization has made caste a central theme for defining differential experience and impact. Its research reports, policy papers, and campaign materials cover different regions and Dalit communities to show social–spatial patterns of production, and the impact of climate change. Connecting the dots between human rights, social justice, and economic injustices on Dalits, its advocacy and campaign efforts concentrate on socio-ecological injustices. For example, it focusses on how climate change arenas like drought, famine, and cyclone impact access and use of natural resources in Dalit localities. To take another example, they highlight how migration to urban areas became a forced but favoured adaptation strategy. Migration may be an accumulative strategy and some rural agricultural labourers may, in the aftermath of climate change-induced hazard, use it to shift from the agricultural to the non-agricultural sector. However, most people end up doing casual labour in urban fringes, often amidst hazardous environments.

Lee Macqueen explains NCDHR's unique work in areas of disaster, Dalits, and climate change.[59] Locational, social, and economic vulnerabilities place a greater strain on Dalits' adaptive capacity to climate change, and ability to deal with shocks, stresses, and change. Their pre-existing vulnerabilities are compounded in disasters, which the country is increasingly susceptible to, in the context of climate change. Two initial ground research works of the organization – the 2004 Tsunami in Tamil Nadu and the 2007 floods in Bihar – became milestones in bringing out realities of Dalit discrimination and exclusion during disasters, and created some ground to develop advocacy and campaign on disaster policy in India.[60] This was followed by more research in the states of Andhra Pradesh, Assam, Karnataka, Maharashtra, Madhya Pradesh, and Uttar Pradesh.

The organisation realized that the institutional mechanisms for disaster management, that is, National Disaster Management Policy, National Disaster Management Act 2005, and related guidelines, do not recognize caste-induced vulnerabilities (except a passing reference to caste in the DM Act Chapter XI: Miscellaneous 6). Casualties and damages, loss of properties, infrastructures, environment, essential services, and means of livelihood on such a large scale are beyond the normal capacity of affected Dalit communities to cope with. Unlike indigenous communities, the state neither recognizes Dalit communities as local minority communities nor acknowledges their contribution to biodiversity conservation. Any policy or programme that does not include Dalit communities among a biodiversity-dependent set of communities remains highly subcritical to adopting adaptation and mitigation measures. Case studies by NCDHR show that Dalit communities remain excluded in climate change adaptation programmes and policies. They thus conclude:

> There is a need for proper state support to develop peoples' adaptation mechanism and support livelihood diversification strategies. Further, the differentiated social impacts of climate change based on gender, caste, class, ethnicity etc., need to be understood better. Disaster risk reduction and climate change adaptation need to inform the social protection policy so the poor Dalits and other marginalized sections can be shielded from shocks and risks owing to climate change and their livelihoods protected effectively.[61]

NCDHR's core mandate of 'global Dalit rights' also involves developing a perspective on climate-related lives of Dalit-like people globally and connecting advocacy work to international arenas. Forging partnerships with various organizations at different places, while highlighting Dalit identity as a key to analysing the uneven impact of climate change, makes the wider world sensitive to the organization's perspective. NCDHR, along with the International Dalit Solidarity Network, has conceptualized a 'Draft Framework for International Humanitarian Stakeholders for Addressing Caste based Discrimination (CBD) in Disaster Response (DR)'. According to Lee Macqueen, the draft is based on a past analysis of Indian experiences of caste-based discrimination in disaster response and DRR, as well as on some good practices and actions that facilitate socio-economic inclusion of Dalits in climate change measures. It also includes several forward steps to address an international community to conceptualize the intersectional ability of Dalits, caste, and climate change, and operationalize some concrete measures for change.[62]

NCDHR has perceptively noticed the high moments of weather and climate, heat and temperature, and wishes to recognize caste-based disproportionate impact of climate change on Dalits at multiple levels. Amidst extreme climate occurrences, they also ground their work on the everyday, invisible, and subtler climate effects on their social and political community. Importantly, they have evolved partnerships with various organizations. Subtler effects include aggregates of the human and the natural, are caste-induced, and are specific to the Dalit community. NCDHR disaggregates the politics of climate justice at the local level. An articulation of Dalit climate justice, capacity building, and community mobilization around certain themes does not preclude an analysis of complexities of climate change politics, and the development of research, policy, and advocacy work at a national and international level. This initiative, research, and capacity building by a Dalit organization has proved to be critical in building a caste- and Dalit-based knowledge of climate change.

A sociological and economic narrative of climate change reveals the slow, subtle, and everyday environmental experiences and processes that are embedded in caste. Dalits' everyday adaptation to seasonal and intermittent environmental changes does not differentiate between weather and climate. The experience of local weather remains concrete and has no ambiguity of the broader climate. The case study of the brick kiln industry of Haryana highlights quantifiable accounts of production, technology, and emission, which are implicitly tied with caste and power relations. While climate is changing, it is lodged in deep structures and ideologies of caste. The climatic scenario, whether spatial, temporal, or measurable, is not neutral, and there are winners and losers in its changing situation. Within the broad umbrella of climate change, its intersections with caste and Dalits, just like gender, need to be clearly recognized and stated. Climate justice derives its strength from a basic consensus in recognizing the distinct and different nature of deprivation and discrimination of people, in this case of Dalits. Climate experts and activists have made an important contribution by adopting a framework of rights, justice, and inequality. At the same time, anti-caste writers and Dalit organizations have significantly enriched our understanding of climate change, and its concepts, methods, and models. They have defined and moulded their politics, mobilization strategies, research, advocacy, and partnerships for capacity building of Dalit communities, and shown why and how caste matters for grasping climate change.

Caste and racial capitalism have permeated various eras and continents, exerting their influence on Earth's systems. Throughout history, Dalit, black, bonded, and slave labourers have been exploited, traded, and even killed in the pursuit of transforming the world. In the present day, Afro and non-Afro descendent people, along with indigenous communities, persist in their battles against environmental racism and casteism, fighting for their rights. The forthcoming chapter aims to showcase the historical and ongoing connections between Dalit and black ecologies, emphasizing their role in environmental stewardship. These valuable insights can serve as a wellspring of knowledge, enabling the creation of archives that will inspire future struggles.

Notes

1. For details, see 'A Dalit View on Climate Change', International Dalit Solidarity Network, 17 December 2009, https://idsn.org/a-dalit-view-on-climate-change/, accessed on 3 February 2024; 'Upper-Caste Farmers Grow Money, We Grow Food', https://idsn.org/resources/case-stories/upper-caste-farmers-grow-money-we-grow-food/, accessed on 3 February 2024.
2. Interview with Ashok Bharti, 20 May 2021. Also, 'Dalits on Dignity March in New Delhi', International Dalit Solidarity Network, 9 December 2009, https://idsn.org/dalits-on-dignity-march-in-new-delhi/, accessed on 3 February 2024.

3. There are several articles and reports on the subject. For example: Sumedha Pal, 'Overlooked Correlation between Climate Change and Social Exclusion', *NewsClick*, 11 July 2019, https://www.newsclick.in/overlooked-correlation-climate-change-social-exclusion, accessed on 3 February 2024; Pranav Prakash, 'The Hidden Casteism of Climate Change Reporting in India', *The Quint*, 27 October 2016, https://www.thequint.com/news/environment/the-hidden-casteism-in-climate-change-media-reporting-in-india-dalit-agriculture, accessed on 3 February 2024; Pradnya Mangala, 'Climate Justice in India: A Critical Overview', *Round Table India*, 6 October 2019, https://roundtableindia.co.in/index.php?option=com_content&view=article&id=9734:climate-justice-in-india-a-critical-overview&catid=119:feature&Itemid=132, accessed on 3 February 2024.

4. Navroz K. Dubash, ed., *Handbook of Climate Change and India: Development, Politics and Governance* (Oxon: Earthscan, 2012); Praful Bidwai, *The Politics of Climate Change and the Global Crisis: Mortgaging Our Futures* (New Delhi: Orient BlackSwan, 2012).

5. For example: Navroz K. Dubash, ed., *India in a Warming World: Integrating Climate Change and Development* (New Delhi: Oxford University Press, 2019); S. S. Jeevan, *Climate Change Now: The Story of Carbon Colonisation* (New Delhi: Centre for Science and Environment, 2018); Amitav Ghosh, *The Great Derangement: Climate Change and the Unthinkable* (Chicago: University of Chicago Press, 2016).

6. E. Somanathan and Rohini Somanathan, 'Climate Change: Challenges Facing India's Poor', *Economic and Political Weekly* 44, no. 31 (1–7 August 2009); K. N. Ninan, 'Climate Change and Rural Poverty Levels in India', *Economic and Political Weekly* 54, no. 2 (12 January 2019).

7. Government of India, 'The National Action Plan on Climate Change', 4, https://envt.kerala.gov.in/wp-content/uploads/2019/05/National-Action-Plan-on-Climate-Change.pdf, accessed on 3 February 2024.

8. Narendra Modi, 'PM's Statement at the United Nations Summit for the Adoption of Post-2015 Development Agenda', 25 September 2015, https://www.narendramodi.in/text-of-pm-s-statement-at-the-united-nations-summit-for-the-adoption-of-post-2015-development-agenda-332923, accessed on 3 February 2024.

9. For example, see the work of Climate Action Network South Asia (CANSA), which is a coalition of around 300 civil society organizations, working in eight South Asian countries to promote government and individual action, to limit human-induced climate change, https://cansouthasia.net/about-cansa/, accessed on 3 February 2024.

10. There are rare exceptions, where social constructions of caste have been taken into consideration in analysing the adaptive capacity to climate change. For example: Annette Lof, *More than Meets the Eye? Exploring How Social Constructions Impact Adaptive Capacity to Climate Change* (CTM Stockholm University: Centre for Transdisciplinary Environmental Research, 2006); Manohara Khadka, Golam Rasul, Lynn Bennett, Shahriar M Wahid, and Jean-Yves Gerlitz, 'Gender and Social Equity in Climate Change Adaptation in the Koshi Basin: An Analysis for Action', in *Handbook of Climate Change Adaptation*, ed. Walter Leal Filho (Berlin: Springer, 2015); Lindsey Jones, *Overcoming Social Barriers to Adaptation* (London: The Overseas Development Institute, 2010); Srilata Sircar, 'Reimagining Climate Justice as Caste Justice', in *Climate Justice in India*, ed. Prakash Kashwan (UK: Cambridge University Press, 2022), 162–81.

11. Dadasaheb Tandale, 'Caste, Economic Inequality and Climate Justice in India', in *Human Rights and Economic Inequalities,* ed. Gillian Macnaughton, Diane F. Frey, and Catherine F. Frey (Cambridge: Cambridge University Press, 2021), 217–44.

12. Abhay Kumar, Aniruddha Dekha, and Rajat Sinha, *Rural Housing in India: Status and Policy Challenges* (New Delhi: Lokashraya Foundation, 2016).

13. Climate Change Programme, *Climate Change and Agriculture in India* (New Delhi: Ministry of Science and Technology, Government of India, 2016).

14. There are many publications signifying this trend. For example: Sunita Narain, Prodipto Ghosh, N. C. Saxena, Jyoti Parikh, and Preeti Soni, *Climate Change: Perspectives from India* (Delhi: UNDP India, 2009); David Michel and Amit Pandya, eds., *Indian Climate Policy: Choices and Challenges* (Washington, DC: The Henry L. Stimson Center, 2009).

15. There are several publications of this kind. To name a few: M. Balasubramanian, Manjunatha Munishamappa, Remadevi O. K., Vinayakumar K. Hombalegowda, and Rittu Kakkar, *Climate Change and Its Impact on Vulnerable Communities: A Case Study of Karnataka* (Bangalore: Institute for Social and Economic Change, 2019); Kelly Bridges, 'Climate Change, Scheduled Caste, and Scheduled Tribes: Analyzing Socioeconomic and Climate Change Vulnerabilities Amongst Female Farmers in Rural Madhya Pradesh', *In Situ* 1, no. 1 (2016); Muniyandi Balasubramanian, 'Climate Change, Famine, and Low-income Communities Challenge Sustainable Development Goals', *The Lancet* 2, no. 10 (2018).

16. NAPCC, https://envt.kerala.gov.in/wp-content/uploads/2019/05/National-Action-Plan-on-Climate-Change.pdf, accessed on 3 February 2024.

17. There are several initiatives and reports on this. For example: Chintan, *Cooling Agents: An Examination of the Role of the Informal Recycling Sector in Mitigating Climate Change* (New Delhi: Safai Sena, The Advocacy Project, Chintan, 2009); Chintan, *Wastepickers: Delhi's Forgotten Environmentalists?* (New Delhi: Chintan, 2019); Prabha Khosla and Bharati Chaturvedi, 'Mitigation of Greenhouse Gases (CHGs) by Informal Waste Recyclers in Delhi, India', in *Gender and Climate Change: An Introduction,* ed. Irene Dankelman (London: Earthscan, 2010).

18. The aforementioned studies and many others do not even mention caste- or Dalit-related issues.

19. Mark Twain, *Following the Equator and Anti-imperialist Essays* (New York: Oxford University Press, 1996), 109.

20. Urmila Pawar, *The Weave of My Life: A Dalit Woman's Memoirs*, trans. Maya Pandit (New York: Columbia University Press, 2009).

21. Ibid., 45.

22. Vasant Moon, *Growing Up Untouchable in India: A Dalit Autobiography*, trans. [from Marathi] Gail Omvedt (Maryland: Rowman and Littlefield Publishers, 2001). Gail Omvedt writes in her introduction that though she herself has experienced it, she has never found such a compelling description of Nagpur's heat.

23. There are many important indicators of climate change. I am only focusing on weather and climate here.

24. In this context, several Dalit autobiographies can be read. For example: Om Prakash Valmiki, *Joothan: A Dalit's Life*, trans. Arun Prabha Mukherjee (Kolkata: Samya, 2007); Sharankumar Limbale, *The Outcaste: Akkarmashi*, trans. Santosh Bhoomkar (New Delhi: Oxford University Press, 2003); Manoranjan Byapari, *Interrogating My Chandal Life: An Autobiography of a Dalit*, trans. Sipra Mukherjee (New Delhi: Sage, 2018); Adwaita Mallabarman, *A River Called Titash*, trans. Kalpana Bardhan (Berkeley: University of California Press, 1993).

25. Pawar, *The Weave of My Life*, 34.

26. Ibid., 43.

27. Quoted in Pal, 'Overlooked Correlation between Climate Change and Social Exclusion', 11 July 2019.

28. The fieldwork was conducted in the brick kilns of Passour, Badli, Kablana, Khungai, and Kanonda villages of Jhajjar district in June–July 2018.

29. For details on Jhajjar District, see Directorate of Census Operations, *District Census Handbook Jhajjar: Village and Town Wise Primary Census Abstract (PCA)* (Chandigarh: Government of India, 2011).

30. For details, see Government of India, Ministry of MSME, *Brief Industrial Profile of Jhajjar District* (Karnal: MSME-Development Institute, n.d.). The report gives information on a steady increase in the number of registered units from 1991–92 to 2010–11. The total number of industrial units were 2,500 in 2010–11, of which 1,849 were registered.

31. Several districts of Haryana – Bhiwani, Ballabhgarh, Jind, Karnal, Palwal, Panipat, Sonipat, Rewari, and Nuh – have brick kilns in big numbers.

32. SAARC Energy Centre, *Evaluating Energy Conservation Potential of Brick Production in India* (Islamabad: SAARC Energy Centre, 2013).

33. Suneet Chopra, 'Bondage in a Green Revolution Area: A Study of Muzaffarnagar Brick-Kiln Workers', *Social Scientist* 10, no. 3 (1982): 38–55; A. Dharmalingam, 'Conditions of Brick Workers in a South Indian Village', *Economic and Political Weekly* 30, no. 47 (1995): 3014–18; Government of India, *Shramshakti: Report of the National Commission on Self-Employed Women and Women in the Informal Sector* (New Delhi, 1988); Leela Gulati, 'Female Labour in the Unorganised Sector: Profile of a Brick Worker', *Economic and Political Weekly* 14, no. 16 (1979): 744–52; Leela Gulati and Mitu Gulati, 'Female Labour in the Unorganised Sector: The Brick Worker Revisited', *Economic and Political Weekly* 32, no. 18 (1997): 968–71; D. P. Singh, 'Women Workers in the Brick Kiln Industry in Haryana, India', *Indian Journal of Gender Studies* 12, no. 1 (2005): 83–96.

34. Karin Lundgren-Kownacki, Siri M. Kjellberg, Pernille Gooch, Marwa Dabaieh, Latha Anandh, and Vidhya Venugopal, 'Climate Change-Induced Heat Risks for Migrant Populations Working at Brick Kilns in India: A Transdisciplinary Approach', *International Journal of Biometeorology* 62, no. 3 (2018): 347–58.

35. Brooke, *The Danger of Heat Stress in India's Brick Kilns* (London: Brooke Action for Working Horses and Donkeys), https://www.thebrooke.org/our-work/india/danger-heat-stress-indias-brick-kilns, accessed on 3 February 2024. Also see Brooke, *Brick by Brick: Environment, Human Labour and Animal Welfare* (London: Brooke, 2017).

36. For example: Shakti, *A Roadmap for Cleaner Brick Production in India* (Delhi: Shakti Sustainable Energy Foundation, 2012); Andrew Eil, Jie Li, Prajwal Baral, and Eri Saikawa, *Dirty Stacks, High Stakes: An Overview of Brick Sector in South Asia* (Washington, DC: The World Bank, 2020).

37. Information given by the National Bank for Agriculture and Rural Development (NABARD) office, Jhajjar. According to the office, the gross irrigated area in the district was 195,028 hectares in 2016–17, which was 80 per cent of the gross cropped area.

38. Jat farmers in Haryana have also been going through the impact of agrarian crisis since the past few decades, leading to several economic and social developments in the state. There are many reports and articles on this: Jitendra and Rajat Ghai, 'Going Backward', *Down to Earth*, 31 March 2016, https://www.downtoearth.org.in/coverage/agriculture/going-backward-53194, accessed on 3 February 2024.

39. Government of Haryana, Department of Revenue and Disaster Management, *District Disaster Management Plan Jhajjar* (Jhajjar: Haryana Institute of Public Management and District Administration, 2018).

40. A case study of heat stress faced by brick kilns labourers near Chennai confirms that labourers are exposed to extreme outdoor temperatures (maximum ranges between 40°C and 45°C in the hot months), and high radiant heat from brick kiln furnaces over many hours. The heat analysis was divided into: (*a*) environmental heat measurements; (*b*) internal heat production; (*c*) work clothing trapping heat; and (*d*) self-reported heat stress. See Karin Lundgren-Kownacki et al., 'Climate Change-Induced Heat Risks'.

41. There are several works on occupational health and safety issues of brick kiln workers. For example: Sunita Kumari, 'Occupational Health of Brick Workers of India', *International Journal of Health Sciences and Research* 183, no. 8 (2018): 183–89; Rufiat N. Kazi and Mangala M. Bote, 'A Cross Sectional Study to Determine the Health Profile of Brick Kiln Workers', *International Journal of Community Medicine and Public Health* 6, no. 12 (2019): 5135–41.

42. It is worth mentioning that in 2019, the National Green Tribunal ordered to close down the brick kilns operating in the National Capital Region districts of Haryana, Uttar Pradesh, and Rajasthan states because the brick kilns fuels were contributing to the worsening PM load in Delhi–NCR air quality, beyond their carrying capacity. The Central Pollution Control Board was asked to carry out a study on the location to determine the emissions coming out of brick kilns and how and when alternative 'Zig-Zag' technology can be applied to the polluting brick kilns. For details: https://greentribunal.gov.in/sites/default/files/all_documents/Report%20in%20O.A.%20No.1016%20of%202019.pdf.

43. D. R. Pattanaik, M. Mohapatra, A. K. Srivastava, and Arun Kumar, 'Heat Wave over India during Summer 2015: An Assessment of Real Time Extended Range Forecast', *Meterology and Atmospheric Physics* 129, no. 4 (2017): 375–93. According to the Centre for Research on the Epidemiology of Disasters, Brussels, it was the fifth deadliest heatwave in the world and the second deadliest in India: Subodh Varma,

'You're Experiencing World's 5th Deadliest Heatwave Ever', *Times of India*, 31 May 2015, https://timesofindia.indiatimes.com/india/youre-experiencing-worlds-5th-deadliest-heatwave-ever/articleshow/47485972.cms, accessed on 3 February 2024.

44. Government of Haryana, *Haryana State Action Plan on Climate Change* (Chandigarh, 2011).

45. Nisha Onta and Bernadette P. Resurreccion, 'The Role of Gender and Caste in Climate Adaptation Strategies in Nepal: Emerging Change and Persistent Inequalities in the Far-Western Region', *Mountain Research and Development* 31, no. 4 (2011): 351–56.

46. Nicole Detraz, 'Threats or Vulnerabilities? Assessing the Link between Climate Change and Security', *Global Environmental Politics* 11, no. 3 (2011): 104–20.

47. Sherilyn MacGregor, 'Moving beyond Impacts: More Answers to the "Gender and Climate Change" Questions', in *Understanding Climate Change through Gender Relations*, ed. Susan Buckingham and Virginie Le Masson (London: Routledge, 2017), 18.

48. Robert D. Bullard and Beverly Wright, eds., *Race, Place and Environmental Justice after Hurricane Katrina* (Boulder: Westview Press, 2009).

49. Ashwini Deshpande, *The Grammar of Caste: Economic Discrimination in Contemporary India* (New Delhi: Oxford University Press, 2011).

50. Barbara Harriss-White, *India's Market Economy* (Gurgaon: Three Essays, 2005), xi.

51. Ibid., 136.

52. Aseem Prakash, *Dalit Capital: State, Markets and Civil Society in Urban India* (New Delhi: Routledge, 2015), 18.

53. Harish Damodaran, *India's New Capitalists: Caste, Business, and Industry in a Modern Nation* (New York: Palgrave Macmillan, 2008).

54. Aaron M. McCright and Riley E. Dunlap, 'Cool Dudes: The Denial of Climate Change among Conservative White Males in the United States', *Global Environmental Change* 21, no. 4 (2011): 1163–72.

55. Cynthia Enloe, *Seriously! Investigating Crashes and Crises as if Women Mattered* (Berkeley: University of California Press, 2013).

56. There are a large number of publications on climate justice. For example: Jeremy Moss, ed., *Climate Change and Social Justice* (Victoria: Melbourne University Press, 2009); Tracey Skillington, *Climate Justice and Human Rights* (New York: Palgrave Macmillan, 2017); Clare Heyward and Dominic Roser, eds., *Climate Justice in a Non-Ideal World* (UK: Oxford University Press, 2016).

57. For details, see 'NCDHR 20 Years Journey', http://www.ncdhr.org.in/, accessed on 3 February 2024.

58. Interview with Paul Divakar, 25 June 2021. For further NCDHR perspectives on climate change, see National Dalit Watch of National Campaign on Dalit Human Rights and Society for Promotion of Wastelands Development, *Impact of Climate Change on Life and Livelihood of Dalits: An Exploratory Study from Disaster Risk Reduction Lens* (Delhi: National Dalit Watch of National Campaign on Dalit Human Rights and Society for Promotion of Wastelands Development, 2013).

59. Interview with Lee Macqueen, Delhi, 25 June 2021.

60. NCDHR has prepared different briefs and reports on states like *Inclusion Assessments, Cyclone Fani: Tracking Inclusion of Dalits, Adivasis, Minorities and other Marginalised Communities in the Disaster Response, The Extent of Inclusion of Dalit and Adivasi Communities in the Post-disaster Response in Kerala 2018.*

61. NDW of NCHDR and SPWD, *Impact of Climate Change on Life and Livelihood of Dalits*, 15.

62. For details, see National Dalit Watch of National Campaign on Dalit Human Rights, *Addressing Caste Discrimination in Humanitarian Response* (New Delhi: NCDHR, 2011).

9

Caste and Race*

Environment Justice and Intersections of Dalit–Black Ecologies

On 15 August 1973, the twenty-sixth anniversary of Indian Independence, Dalit Panthers, an organization founded by young Dalits in 1972 to evoke ideas of blackness and black power for representing Dalit politics, took out a march on Mumbai's streets. The march was called the Black Independence Day (Kala Swatantrya Din), where a Dalit Panthers Manifesto was released. Questioning the dominant narratives of independence, democracy, government, and political parties in India, the manifesto identified three 'burning questions of Dalits today': '1) food, clothing, shelter; 2) employment, land, untouchability; 3) social and physical injustice'. Key programmes, carried on under this manifesto, have extensively focused on Dalit access to natural resources: 'The question of landlessness of the dalit peasants must be resolved'; 'Dalits must be allowed to draw water from public wells'; 'Dalits must live, not outside the village in a separate settlement, but in the village itself'; and 'all means of production must belong to the Dalits'.[1]

Political links between the Dalit Panthers and the Black Panthers, common legacies of B. R. Ambedkar and W. E. B. Du Bois, and numerous initiatives and academic writings by organizations, activists, and academicians have aligned anti-caste struggles of untouchables and low-caste people in India with anti-racist articulations of African Americans in the United States.[2] Amid this long tradition of intersections between anti-casteist and anti-racist positions,[3] Dalit–black solidarities have encompassed questions of social and economic inequalities, civil and political rights, possibilities of democracy, development, gender, empowerment, and formation of organizations of the oppressed.[4]

* Originally published as Mukul Sharma, 'Caste, Environment Justice, and Intersectionality of Dalit-Black Ecologies', *Environment and Society: Advances in Research* 13 (2022): 78–97. Reprinted with permission.

In history and economics, too, caste and race, Dalits and black people, and the parallel connections and common grounds between the two have been widely analysed. Caste and race have often been considered interchangeable, in terms of their overwhelming potential for damaging human society.[5]

However, the age of civil and political rights for India's Dalits and America's black people is also a time of environmental inequalities and social and ecological movements in both communities and continents. Yet, the search for shared histories and struggles of Dalits and black people have not entered the unquiet world of environmental struggles. This disjoint between the dynamic and long history of Dalit–black solidarity, and the near absence of an overlapping environmental discourse, is something of a historical puzzle. Amid this background, this chapter analyses how ecologies of Dalits and black people are similar and/or different. It focuses on a comparative structuring logic between environmental racism and environmental casteism, as both are forms of racialization and spatialization, and both fundamentally pivot on embodied and lived experiences. The chapter attempts to think of dynamic intersections between Dalit and black environmentalism for future work in academia and activism.

I hinge on four broad themes. The first is the articulation and expansion of environmental justice movements of colour and low-income communities in the United States, and how these can be useful in framing Dalit responses, and for conceptualizing their environmental struggles against protean forms of eco-casteism. Based on a broad literature review, I outline the dynamic, heterogeneous, and plural discourses of environmental justice, and its specific histories among African Americans. As has been stated, 'In this sense, the discourse of environmental justice may be seen as a unifying process, bringing together diverse situations and sharing understandings and experiences.'[6] Second, I draw on certain key literary writings to study some historical and contemporary ecological struggles of Dalits. Deploying three case studies, I demonstrate how environmental articulations under the rubric of civil rights developed into significant struggles over Dalits' access, occupation, and rights in the natural and physical environment, and how in the process, themes of social and environmental justice appeared on the forefront. Through contemporary cultural, social, and political assertions on land, water, and commons, Dalits are opening up a new ecological universe, which is often outside the dominant discursive framework. In the third rubric, I draw on a subset of literature by anti-caste and Dalit writers to capture Dalit caste conceptions of environment, and the making of Dalit ecologies. Purity and pollution of body, touch, taste, space, place, and people are key markers of caste, creating essential qualities and differences within and outside of naturescapes. In Dalit relations to environment, they articulate certain notions of nature that go beyond physicality, possession, and distribution of resources as environmental justice discourses have done in the past. The language of experience, feeling, humiliation, and dignity, ubiquitously used in Dalit movements, also gets integrated into conceptions of environmental justice. Finally, based on the parallels drawn between environmental casteism and environmental racism, and simultaneous struggles against them, I contemplate how we can conceive of a new counter-archive, where a rich repertoire of Dalit and black ecologies – their cultures and memories, stories and struggles, systems of knowledge and technological skills – can talk to each other. Cornel West and Suraj Yengde[7] call this an exercise in connecting histories, building power, fusing something, and creating 'a new historical archive' to strengthen a long tradition of Dalit–black solidarity. Inspired in part by black feminist and critical race and caste studies,[8] my approach draws from an understanding of intersectionality as a conceptual tool. Intersectionality underlines how identities and systems intertwine to produce individual and relational experiences. As Patricia Collins states: 'As opposed to examining gender, race, class and nation as separate systems of oppression, intersectionality explores how these systems mutually construct one another' and that 'certain ideas and practices surface repeatedly across multiple systems of oppression'.[9] The intersections between Dalit and black ecologies can help in creating a robust archive of environmental justice.

Environmental Justice in the Historical and Global Context

The evolution of the environmental justice movement among black people in the United States in the 1980s has been widely researched. Its historical roots can be traced back to a fundamental reorganization of space in America after World War II, and the state-sanctioned racial discrimination and segregation of people of colour, in terms of housing, workplace, location of dangerous and polluting-emitting factories, and disposal of toxic wastes.[10] Robert Bullard[11] brings out how toxic dumping, municipal waste facility siting, and discriminatory land-use decisions in areas of colour and poor communities were a central concern of the black activists in the early environmental justice movement. Warren County, North Carolina, became a symbol of the environmental justice movement when a large number of African American protestors laid down on the highways for weeks, and were arrested in big numbers, for blocking dump trucks carrying dirt laced with high concentrations of cancerous polychlorinated biphenyls toxic chemicals from entering waste landfills. Whispering Pines Sanitary Landfill and protests by the Houston Northwood Manor subdivision residents, outcry over the prevalence of toxic chemicals in Louisiana's Cancer Alley, and many such instances in black localities prompted research, publications, and activism around disproportionate siting of hazards, toxic, and industrial polluting facilities in black and brown communities. Simultaneously, farm workers' struggles against the use of dangerous pesticides and their impact on health and industrial action against environmental hazards at the workplace by industrial labour organizations, such as the Oil, Chemical, and Atomic Workers Union; and concerns around occupational health and working conditions of sanitation workers appeared prominently on the agenda of working class and environmental justice grassroots groups. During this period, working-class environmental groups were formed to reduce pollution in the community. Focus was on air and water pollution, factory emissions, and improved sanitation (illegal dumping, garbage removal).[12] Researchers like Jedediah Britton-Purdy adopt a historical lens to describe 'a long environmental-justice movement', of more than a century, in the US. Several examples, like the Wilderness Act 1964, and a great victory for a long political drive to preserve 100 million acres of public land, or the proposal of Miners for Democracy, which briefly took over the United Mine Workers of America in the early 1970s to enforce safety regulations and environmental principles in the workplace, or the Earth Day 1970, with the largest mass mobilization in American history, demonstrated that historically, environmentalism was also a social justice movement. At the same time, Purdy pinpoints three central criticisms made by environmental justice scholars and activists regarding mainstream environmentalism. First, it does not recognize the distribution of environmental harms and benefits along the lines of poverty and race. Second, environmental justice questions the mainstream environmental idea of what environmental problems are in the first place. Third, mainstream environmentalism overemphasizes elite forms of advocacy and has less space for people's mobilization.[13]

Critiques have also been offered against the 'whiteness of the environmental movement'[14] and the exclusive nature of major environmental organizations (Letter to Big Ten Environmental Groups, 16 March 1990). Historians have pointed out that the first national People of Color Environmental Leadership

Summit, held in Washington, DC, in October 1991, represented a historic moment in the environmental justice movement. Among the delegates of African American, Native American, Asian American, and Latino origin, Dana Alston, an African American environmentalist, delivered her now-famous speech, 'The environment, for us, is where we live, where we work and where we play,' which was a shift from mainstream environmentalism.[15]

Seventeen 'Principles of Environmental Justice' adopted at the summit, and a defining document of the environmental justice movement, emphasized the importance of racial justice for people of colour. Simultaneously, it affirmed 'the fundamental right to political, economic, cultural and environmental self-determination of all peoples' (Adopted at the First National People of Color Environmental Leadership Summit Washington, DC, 24–27 October 1991). Thereafter, conceptual and organizational trajectories of the environmental justice movement multiplied manifold, though the context of injustice remained central, and race continued as its key component. Alongside, the movement reflected the pluralities of justice discourse, and encompassed civil and human rights, injustices, inequalities, exclusion, and victimization; politics of place and spatial dimension of justice; and recognition of experience, difference, diversity, and participation of black ecology.[16]

Social justice has found frequent mention in the vocabulary of environmental justice.[17] Key elements of social justice have been defined in terms of environmental inequalities, exclusion, discrimination, harms, victimization, distribution, access and rights, and their relevance to specific social groups.[18] David Harvey concretizes 'the environment of justice'[19] by positing nature as internal to society and all ecological projects as political and social projects. Nature would not have existed in its present form had humans not been mixing their labour with the land all along. For him, the present and the future of nature – the new Earth and the new humanity – should be understood through labour and spatial and social change.[20] According to William Cronon, from the perspective of justice, the central question is, 'Whose Nature?' He suggests that mainstream environmental politics and ethics should frontally acknowledge the deeply troubling truth that 'nature' – which they seek to understand and protect – is not 'out there' but is produced in experiences, ideas, and imaginations 'right here'.[21] There are many 'human versions of nature' and they will be jostling and contesting against each other. A social justice framework includes dignity and respect, protection of human rights, social equality, and economic egalitarianism, where each person has the same rights, opportunities, and services as all other people. It marks active participation in sociopolitical institutions and decision-making, which affects individuals, groups, and collectivities, of which they are a part. Four concrete propositions have been placed to make eco-justice more relevant: justice as an active process, justice through maximizing liberty, justice dealing with issues holistically, and justice's temporal and spatial dimensions.[22]

Environmental justice activism has been grounded locally; at the same time, it has continued to evolve globally.[23] Not only has its initial perspective been anchored in bourgeoning thinking about the interconnectedness of environmental justice with 'different cultural and political histories' and 'the rejection of any form of racism, discrimination and oppression (Second People of Color Environmental Leadership Summit) but it has also had good traction in different continents around broader concepts of justice. Horizontal and vertical globalizing of environmental justice has found echoes in different countries and continents. Going beyond national borders, its concerns have encompassed trade agreements, transfer of wastes, climate change, and the Rights of Nature.[24] Drawing strong connections between environmental and social issues, environmental justice provides an opportunity for developing diverse politics and broad-based coalitions all around.[25] Environmental justice research and publications have flourished through several country-specific studies.[26] While black people have laid the foundations of environmental justice by questioning the dominant paradigm of environmentalism, the globalism of the movement has raised some pertinent questions about its conceptual trajectories. However, the moot question is whether 'the popular understanding of environmental justice is based on too narrow a view of "environment" and too narrow a view of "justice"'.[27] Does the concept encompass realities of Africa, Asia, and Latin America, and articulate a wide range of specific social, cultural, economic, and environmental issues? Further, how should the global discourse on environmental justice, generated by black Americans and others, take note of the role of caste in organizing social and environmental relations, and the complex ways in which the Indian caste system creates hierarchical power structures, and works through other centres of power, to naturalize and organize environmental inequalities?

Even the 'new pluralism' of environmental justice, which offers ways to include diversity, difference, recognition, and participation, for unpacking histories and geographies of exclusion, makes no reference to caste and Dalits. Indian environmental discourse, otherwise sensitive to the issue of environmental justice, too, falls short here. For example, 'A History of Environmental Justice in India' succinctly traces the history of Indian environmentalism from the lens of social justice. According to the author, the country witnessed the emergence of issues of differential access to natural resources and ecosystem services in the 1970s and 1980s. Thereafter, two other critical environmental justice issues – development-induced displacement and gender-based discrimination in accessing natural resources – were articulated. The Bhopal Gas Disaster of 1984 raised new issues related to the impact of toxic contamination on individuals and communities, where the heavy burden on the poor was also realized. The study claims that the sensibility of environmental justice has been a 'central meme in Indian environmentalism since the 1970s'.[28] However, the criticality and specificity of caste and Dalits in accessing natural resources and ecosystem services has remained on the fringes in such studies. Similarly, discourses of 'environmentalisms of the poor' have been repositories of vocabularies and languages of the rights of the subordinated. Through historical and comparative perspectives, which encompass gender and class, environmental authors have argued that varieties of environmentalism in South Asia 'originate in social conflicts over access to and control over natural resources'[29] – for example, conflicts between peasants and industry over forest produce; rural and urban populations over water and energy; or struggles of the poor against corporates, markets, and the state to retain their control over natural resources. Yet, the authors treat struggles for environmental justice by various social groups as separate from Dalit issues, and they elide or subsume the 'caste question'. Consequently, there has been little understanding of how caste intersects with environment to create socio-environmental inequalities in South Asia.

Most Dalit movements, too, have not seriously tried to conceptually intersect the cause of Dalits with environment, in the way black people and other people of colour have. There could be at least two reasons for this. First, some Dalits, like many others (for example, a section of working class, trade union, and left movements), perceive the environment movement as elitist in nature.[30] Second, in a specifically Indian context, being largely concerned with communities (for example, the indigenous forest-inhabiting populations), whose dispossession is tied to modern development and capitalism, environmental movements here have largely focused on the protection of villages, and rural and traditional livelihoods. This has often been anathema for a large section of Dalits, for whom the whole point is precisely to get away from stigmatized traditional occupations, as I demonstrate in my next section on Dalit ecological struggles.

Taking this into account, there has emerged a 'Dalit critique of environmental justice in India',[31] which has also specified the need to evolve new perspectives and priority areas in various movements, which include Dalits and low-caste people. Some stray research efforts have tried to redefine environmental justice in South Asia by including caste and discrimination.[32]

Dalits Ecological Struggles: Where We Live, Where We Work ...

> This water lake of Mahad is a public property. The Touchables of Mahad are so generous that they allow all the people to draw water; even the Muslims and persons of other religions also are free to draw water from there. Not only that, they do not have objection to even animals drinking water from this lake. They allow all animals even those belonging to the Untouchables....

> The Touchables of Mahad are opposing Untouchables in drawing water from Mahad lake not because its water will get polluted or it will vanish in thin air. They oppose it because they do not want to accept that the Untouchables are equal to them.[33]

This was B. R. Ambedkar, launching the Mahad struggle and the burning of the *Manusmriti* in 1927, at the core of which was an assertion of untouchables' right to take water from public waterbodies. When the struggle over Chawdar tank was at its peak, the agitators also publicly burnt the *Manusmriti*. However, this passage or other similar ones have not found a place in the standard chronicles of Indian environmentalism and, even when they do appear, they are read not as an expression of an environmental tradition but simply as discourses on social justice. It is true that Dalits have generally articulated their struggles under the rubric of the 'social' as opposed to the explicitly 'environmental'. At the same time, Dalits are excluded from natural resources, live separately in villages, suffer from landlessness, have restricted access to waterbodies due to purity–pollution taboos, and live in segregated and dirty city spaces, working in 'dirty' occupations. Viewed from this perspective, Dalits have had a long history of social and environmental struggles, which have manifested in regular conflicts against Brahmin and *savarna* (higher caste) domination of eco-space, as well as in assertion and creation of an autonomous Dalit eco-space. Based on Dalit and anti-caste literature, in this section I offer snippets of three such struggles – of the 1920s, 1950–1960s, and 2000s – from different regions of the country, where human and environmental rights overlapped with social and ecological justice and shared a common ground. At the same time, it needs to be noted that Dalit is not a homogeneous category, in terms of labour, occupation, knowledge, and cultural practices. Studying Dalit ecological politics provides us with not only Dalit perspectives but also points to the significant differences in environmental attitudes among them in different regions and reveals the possibilities of debate and varied practices within Dalits.

Let me begin with the Mahad *satyagraha* in western India, which has been researched widely, and described as an epoch-making moment, with far-reaching effects on untouchables, leading to the emergence of a new Dalit politics.[34] Thought-provoking anti-caste interpretations of the Mahad *satyagraha*, however, can be complemented and enriched by employing the lens of environmental justice. This struggle symbolized the ties between untouchability, civil rights, and environmental sensibilities. The centrality of access to natural resources for untouchables, and Hindu religious-based caste inequality and injustices in nature, became a converging point for divergent Dalit political and social traditions.

Hindu Brahmanical scriptures have coloured water with caste. Ideas of ritual purity and pollution, and daily practices and habits of drinking, bathing, fishing, and transportation have been profoundly affected by caste, sanctifying the social order of water. Caste and the Hindu religion have critically come into play when determining Dalits' contact with water, resulting in a tense relationship between water, dominant Hindu discourses, and Dalits. It has been an association of domination on the one hand, and marginalization and exclusion on the other.[35] In his political journey, Ambedkar time and again narrates the tales of water and the Mahad struggle. Located in the Kolaba district of the then-Bombay Presidency, Mahad town had a population of around 8,000, of whom fewer than 400 were untouchables. The Chawdar tank, an old public tank owned by the municipality, was a vast expanse of water, mainly fed by the rains and a few natural springs. The tank lay at the heart of the Hindu quarters and was surrounded by upper-caste Hindu residences. The untouchables had to come to the town for various purposes: shopping, paying government revenues, or performing their duties as village servants. This was the only public tank in the town from which an outsider could get water. Even then, untouchables were barred from fetching its water, causing great hardship. In 1923, the Bombay Legislative Council and the government passed a resolution and issued orders that the untouchables be allowed to use all public water places, wells, and *dharmashalas*, which were built and maintained out of public funds. However, orthodox caste Hindus refused to comply with the order.

It was thus decided to hold a Conference of Untouchables at Mahad, presided over by Ambedkar, on 19–20 March 1927. On 20 March, the Conference decided that, as a collective body, they should go to the Chawdar tank and help the depressed classes in establishing their right to take water from it. The delegates accordingly began to march peacefully towards the Chawdar tank. Ambedkar narrates that the procession marched past and went to the Chawdar tank, and the untouchables, for the first time, drank its water. However, the religious orthodoxy felt threatened, and the priest of the temple next to the tank spread a rumour that the untouchables were also planning to enter the temple. A riot broke out. Caste Hindus attacked Dalits, many of whom were severely wounded. Caste Hindus declared the Chawdar tank to be desecrated by the touch of the untouchables and, soon after, ritually purified its water for their use.

Dalit struggles, however, continued, and another *satyagraha* was started on 25 December 1927, in spite of caste Hindus' aggression in between. The Mahad Municipality revoked its 1924 resolution of opening the tank to the depressed classes, and leaders of orthodox Hindus filed a suit against Ambedkar and others, requesting a temporary injunction. However, in December 1927, thousands of Dalits gathered in Mahad. The mood was to defy the injunction and take water from the tank at whatever cost involved. However, amid several twists and turns, Ambedkar decided to fight the matter in the courts rather than on the streets. On 27 December, a symbolic procession was taken out through the streets of Mahad to the tank, but it did not stop to take water. The court case was finally decided by the Bombay High Court in 1937 in favour of Ambedkar, stating that the caste Hindus' right to exclude untouchables was not based on immemorial custom. The Mahad struggle brought out a complex set of interrelated issues, structures, and aggregates in a caste society – environmental, human, religious, historical, local, legal, political, governmental, and organizational – which were bound together by Dalit resistance and anti-caste struggle.

Let me move to my second example. Based on historical research and oral narratives, Badri Narayan[36] has extensively studied the Nara-Maveshi movement of the 1950s–1960s in north India. Largely unknown till his study, the movement forcefully brought out and challenged the close links between the caste system, occupational segregation, and Dalits. Caste creates a concept of natural and social order, where people, place, occupation, and knowledge are characterized by pollution and ritual cleanliness; where bodies, behaviours, situations, and actions are isolated, 'out of place', and 'untouched' because of deep-down hierarchical boundaries. Casteism often rests on naturalism, where nature is used and abused to provide a body of knowledge, including bonds, locations, and landscapes, for determining individual–collective identities and relationships, in an ecological setting. In such hereditarily ordered categories, Dalits are particularly tied with degrading (polluting) traditional occupations.

Nara-Maveshi signifies two occupational activities, designated traditionally to the Chamar caste: women as midwives cutting the umbilical cord (*nara*) of newborns, and men manually disposing of the corpses and carcasses of dead animals (*maveshi*). Both these occupations have historically been tied with untouchability. Faced with everyday humiliation of their caste-based polluted labour, while also receiving meagre grains and clothes in lieu of wages, Chamar men and women, mainly from Uttar Pradesh and Bihar, decided to stop skinning and tanning dead cattle and cutting newborns' umbilical cords. Narayan takes us through life narratives of several Chamars in villages of Uttar Pradesh to show how they felt dirty and demeaned in their occupation. The movement spread widely, but also invited a range of painful and insulting backlashes from the higher and middle castes of the villages on the Chamars – physical attacks and destruction of homes and property; imposition of 'economic sanctions', including no work on agricultural fields, no loans, no permission to collect firewood from the trees, or to walk through their fields; and prevention of their walking on the roads, drinking water from the wells, or using water for irrigation. Even the local shopkeepers were pressurized not to sell grains to them, while other communities such as washermen and barbers were asked not to render any services to the agitating Chamars.

The movement played a significant role in liberating Chamars from their caste-based 'natural' and social spaces. It expanded in its scope, to include issues of rights and dignity. While most intensive in the 1950s–1960s, the campaign continued in Uttar Pradesh till the 1980s, and attempted to fracture the links between place, occupation, inherited status, and social hierarchy. It brought new perspectives on Dalit meanings of labour and environment.

In my third example, I focus on Dalits' nationwide long march, Bhim Yatra, which was set off on 10 December 2015 by the Safai Karmachari Aandolan (SKA: Movement Against Manual Scavenging), and concluded in Delhi on 14 April 2016, the 125th birth anniversary of Ambedkar, after covering 35,000 kilometres in 125 days, across thirty states. The march included caravans of vehicles, processions, meetings, different groups fanning out in different areas simultaneously, street theatre, and cultural performances. Taking on issues of manual scavenging, dry latrines, sewers, and septic tanks under the slogan of 'Stop Killing Us – Stop Killing Us in Sewer and Septic Tanks', the march demanded that the government 'tender an apology to the scavenger community for the historical injustices and centuries of humiliation of making us manual scavengers' and 'eliminate manual scavenging immediately, without any further delay'. The movement strived to 'break the link imposed by the caste system between birth and the dehumanizing occupation', and to reclaim 'our dignity, equity and human personhood'.[37] The march went around hundreds of villages, cities, colonies, public places, and neighbourhoods. It addressed meetings of Dalits and non-Dalits, performed plays and songs, and organized community dining and overnight stays, to forge a broader understanding and unity. SKA had been engaged in various struggles since 1983 by deploying different campaign strategies: conducting surveys to identify dry latrines, users, and those forced into manual scavenging; filing petitions and complaints with government officials at local, regional, and national levels; educating and sensitizing civil society, especially dry latrine users; filing a Public Interest Litigation in the Supreme Court; and networking with individuals, media, and civil society organizations to form wider solidarity and pressure groups.[38]

Across much of India, the practice of manually cleaning excrement from private and public dry toilets and open drains persists. Manual scavengers are mostly from caste groups customarily relegated to the bottom of the caste hierarchy and confined to livelihood tasks considered as deplorable or deemed too menial by higher-caste groups. Their caste-designated work reinforces the social stigma of being unclean and untouchable and perpetuates widespread discrimination. Since the early 2000s, various movements of manual scavengers have been quite vocal and militant. In 2012–13, thousands of women scavengers organized the Maila Mukti Yatra (Dirt Liberation March) in 200 districts of eighteen states, covering more than 10,000 kilometres, under the banner of Rashtriya Garima Abhiyan (National Campaign for Dignity and Eradication of Manual Scavenging). Liberated women scavengers took a lead to reach out to colonies and houses of those women who were still engaged in the practice and motivated them to leave it completely. The march became aggressive many times, as 'pots were burnt at public places' and 'dry toilets were broken at some places'. The march also made it a point to enter public water sources, parks, and tea and barber shops. It was stated: 'In villages where Dalits were not allowed to wear chappals in non-Dalit colonies, and weren't allowed to take marriage processions, rallies were organized with drum beats, and these were headed by women who used to practice manual scavenging.'[39]

Manual scavenging in India has largely been addressed as an issue of social justice and human rights. However, Dalits have also articulated various labour, health, and environmental concerns associated with it, with a greater focus on sanitation. The bios of humanity as part of nature, with caste as its social determinant, is as critical as access to natural resources. The substance, direction, and rates of change in manual scavenging depend not only on the state of organization and technology but also on the condition of social and physical environment. In a set of various effects on human biology, caste, social, and natural factors remain the leading ones. Dalit struggles on sanitation take note of different kinds of biological reactions to a number of socially conditioned processes.

Some of the fundamental themes of environmental and social justice – Dalits' access, ownership, rights, and participation in land, water, forests, and commons – have appeared frequently in these movements. A large number of environmental conflicts and violence against Dalits in India are found to be related to land, water, forest, and sanitation issues.[40] Infrastructure, real estate, industrialization, and mining have led to new forms of dispossession, displacement, and resistance, where the loss is much greater for Dalits. Recent research shows the prominent role of caste in India's contemporary 'land wars' as caste remains firmly entrenched in various land struggles.[41] Dalits have also asserted their right to water by questioning various tenets of Hindu religion, caste, culture, institutions, and practices, which have prevented them from accessing water due to entrenched notions of untouchability, impurity, and pollutants. Simultaneously, there have been numerous violent incidents perpetrated against Dalit water assertions.[42] The passage of the Forest Rights Act 2006, which was enacted to address the 'historic injustice' to the indigenous people living in the forest region, witnessed Dalit movements to reclaim their rights in forests.[43]

Commons – land, forests, waterbodies, ponds, groves, parks, pavements, streets – have a distinctive valance for Dalits. These are places, spaces, sites, and regions of social conflicts and political protests, domination and resistance, construction and destruction, exclusion and violence. From a Dalit perspective, commons have a multi-layered environmental, economic, and social importance. They have been articulating distinct meanings and imaginations of commons, which have both oppositional and alternative aspects. Dalits have brought to the fore the strange semantic of the word 'common'. Ravikumar, a well-known Dalit writer, notes: 'A "common well" means one from which an untouchable cannot draw water, a "common funeral ground" means a place where the body of an untouchable cannot be cremated, a "common market" is where an untouchable cannot even sit.'[44] Dominant village elites usually control the common lands and water resources, and cannot tolerate the demands of Dalits for new entitlements. The right to equal social space that 'essentially consists of the right to enter and use public space' is the new right claimed by Dalits.[45] Not only in the present but Dalit writers and historians also narrate many instances of struggles by Dalits over public spaces during the colonial period.[46]

These examples demonstrate how, in different historical periods, Dalits have shown an awareness of their physical environment and labour, and the dangers that caste-based social ecology poses to their health, community, and well-being. Their actions have had clear civil right agendas, which also underscore their links with environmental concerns. They have given vent to their environmental imaginations and, in the process, created a collage of Dalit ecologies.

Environmental Racism and Environmental Casteism: Key Convergences and Divergences

Dalit and black ecologies have convergences and divergences, and it is important to assess a few structural similarities and differences between the two, as these can provide a broad road map for future work. For instance, there are some important ecological (political, cosmological, psychic) factors that have profoundly and distinctly shaped Dalit environmental experiences. Dalits can be characterized as a subaltern population, subjected to several millennia of cosmological essentialization *in situ*, that is, being 'of the land' in a primordial sense, and simultaneously abject in the cosmological sense of not having been born of the Purusha, and thus out of human universe. Conversely, the subaltern existential condition of black Americans is different. In the colonial phase of early capitalism, they were uprooted from their African homelands and forced into slavery, enduring the Middle Passage, to become an essentially different people, stripped of personhood, in an alien land. This distinction shapes different kinds of spatial/ecological, economic, and political deracination, and generates distinct experiences of abjection, alienation, violence, despair, and hope, as well as diverse possibilities for resistance, action, and emancipation.

Reflecting on the unique trajectories of Dalit and African American struggles, historian Gyanendra Pandey points to several social divisions among the Dalits themselves. At the same time, Dalit identity has been carved in the course of their sociopolitical struggles, where histories of labour and exploitation, hierarchy and stigma helped in creating a common ground for a new political community. Dalit conversions to Buddhism were also meant to establish a new identity, politics, and culture. In contrast, 'the separate identity of the African American people and culture seems to be in place from their arrival on American shores – or so the legend has it. The experience of slavery, the legal and social barriers against access to basic resources for people of African descent in much of the US for much of its history, the visibility of skin colour, and the discourse of 19th century "science," "civilisation" and "race," have served to establish this as common sense.'[47]

Taking 'racialization' and 'racial capitalism' as key analytical concepts, some notable researchers have talked about caste and Dalits in India as the 'production of racial difference', which is 'reproduced through Hindu nationalist, casteist, and colonial projects that generate tacit and explicit consent for continued violence against racialized others'.[48] According to them, there is a close connection between regional forms of racial difference and forms of global racial capitalism. Racialization in India is calibrated through shifting capitalist political economics, and caste, like race, is inextricably bound with capitalism's dehumanizing impulse. However, capital, industry, development, modernity, and globalization have contradictory implications, with no clear-cut homogeneity of connotations for Indian Dalits. Traditional stigmatized occupations and jobs, with which Dalits were previously associated, had signaled increasing aggression and violence against them. Thus, urbanization, modernity, and development are viewed positively by a number of Dalits. They can aid in opening new opportunities of employment, which can be more emancipatory and materially beneficial. Technological progress particularly attracts the middle-class segments of Dalits. Ambedkar was equally emphatic in his understanding that the transformation of nature by powerful economic and technological forces not only had a living impact on separate components of landscapes but was also closely associated with the possibility of changing society altogether, and with it, its inseparable biosocial organ, namely human and humanity.[49]

At the same time, there are several convergences between Dalit and black struggles in the environmental arena. Dalits' lived and embodied experiences of casteization and spatiality, along with their conception of new commons, land, labour, and environmental rights are some of the pegs that connect environmental casteism and environmental racism. In this section, I focus on the critical relevance of caste studies and anti-caste literature in India, to show the ways in which casteism, social inequality, and untouchability interact with physical and natural forces to create specific forms of environmental domination and exploitation. Anti-caste literature emphasizes that everyday practices, of what constitutes environmental activity and thinking, are structured by an archaeology of untouchability in body, contact, touch, smell, feel, belonging, work, and sociability.

What actually is the environmentality of caste? How does caste demonstrate its environmentality, its social nature, and everyday personal and social experiences? Environment is constituted not just by natural resources but also by a combination of social and physical structures. Caste structures are important components of what constitutes the social and the physical, as caste naturalizes human and social phenomena. Naturalization 'refers to ways of fortifying various social, cultural, economic, or political conventions by presenting them as part of natural order'.[50] According to this view, humans have wrongly considered themselves as *above* nature, whereas they should be viewed as *in* nature, which is rich, permanent, and cultural, and often provides national values to guide human actions. The supremacy of 'natural order' is affirmed in major spheres of caste society – life, labour, livelihood, food, animal, and space – which is often synonymous with a conservative Hindu *savarna* belief.

In their groundbreaking work, Gopal Guru and Sundar Sarukkai[51] explain the many ways in which the natural essentially creates the everyday social of caste. According to them, the conceptions of caste as a natural biological process, casteist constructions of social nature; metaphorical descriptions of caste; other social explanations through images of nature; and natural expressions of domination and authority variously demonstrate the formation of a naturalized social based on hierarchy. Even the natural senses of seeing, touching, smelling, tasting, and hearing have a caste sociality. History, culture, and religion further strengthen the intertwining of naturalism and casteism. For example, the 'naturalized' *varnashrama* or *chaturvarnya* system, in which people were believed to be born with natural characteristics and inclinations towards a particular occupation, forced Dalits to continue with certain 'polluted' occupations.

Naturalized environments, however, are lived environments, where experience is central to building recognition, dignity, respect, and justice, since each can be understood in terms of its potential to create new relationships to natural resources, landscape, and human society. Experience becomes primary to human access, assimilation, and agitation. Several contemporary thinkers have emphasized the importance of a subject's experience in varied historical, political, social, and cultural contexts.[52]

Particularly with regard to casteized nature, any analysis of the nature of justice has to consider the objective and subjective dimensions of environmental experiences. Within the figured world of social nature, 'the environment is associated with the daily smells and sights of blight, along with an awareness of ever-present danger and insult to one's body and to the community. Accompanying these threats are the experiences of other forms of injustice and disregard'.[53] In the specific context of India and Dalits, Sundar Sarukkai outlines three important characteristics of lived experiences: one, the freedom to be a part of an experience; two, the freedom to leave any time if the experience is not satisfactory; and, three, to modify the experience, if necessary, to suit one's needs. However, in Dalit lives, 'lived experience is not about what there is but is about what there is not. *Lived experience is not about freedom of experience but about the lack of freedom in an experience.*'[54]

The production of caste spaces of environmental inequality along multiple axes, and their subjective and objective forms, are intersectional themes. Anti-caste thinkers often articulated that experience, space, and justice matter for the formation of thought and action. Recognizing an overarching influence of space on Dalit lives and thinking, it is argued that in the case of Ambedkar and M. K. Gandhi, space determined the emergence and efficacy of their thought. The language of discrimination, humiliation, and segregation in Ambedkar was a result of his location and space, a social ghetto that was historically produced and reproduced. Gopal Guru explains that experience is subjectively realized but objectively produced through the logic of space. The production of space hinges on the reproduction of space. For a tormentor, space is a certain supporting condition to produce tormenting experiences, which become stable across time in restructured spaces. According to Guru, in Dalit life, spaces have complex, multiple connotations. 'Spatial experience' leads to a language and politics of mobilization of Dalit masses, to radically subvert dominance; 'experimental' space creates a social thought of non-Brahmin thinkers, in opposition to sacred, dominant, closed, and rigid space; 'space as culturally constructed phenomenon' turns untouchable bodies into cultural spaces to rule over and write on them; 'hierarchical spaces' yield different concepts like service, sacrifice, practice, self-respect, labour, dignity, rights, and social justice with different degree of emphasis; and 'material space and social justice' in both sacred and profane locations are sites of intervention where language of dignity and rights takes precedence.[55]

The stated ideals of commons – that they are supposed to be collective and inclusive, capable of supporting people's lives and livelihoods – makes them ideal sites for Dalits, in their quest for equitable distribution of physical and social spaces. For a section of Dalits, in contrast to the traditional and the rural, urban spaces have often symbolized freedom from caste segregation, and sites for entry into the modern. The journey from the village to the city has often been considered by Dalits as a leap into a new world space. Nagraj thus states: 'This idea has all the exodus motifs, including an escape from persecution and a journey towards a Promised Land.'[56] However, in her recent work, Malini Ranganathan notes that in parts of urban India, particularly in Dalit-majority slums, racialization and environmental casteism operate through criminalizing discourses and planning policies, which organize urban spaces and ecologies by containing, disciplining, and evicting Dalits from particular areas of the city.[57]

Some recent research on practising caste in India has opened new ways to describe caste society, sociality, and sociability, which has implications for understanding Dalit environmental justice in theory and practice. Caste is concretely generated through a divide between touching and not touching in body and space relations, thus creating inside and outside divides. Aniket Jaaware analyses the core of caste society in 'Touch' as it constitutes (and is constituted by) touchability/untouchability (a matter usually understood as part of the more general theme of caste) in society.[58] The fundamental characteristics of touch have material–physical and non-physical elements, which include inertia (contact between skin and something else), density (threshold of density to the sense of touch), reality (non-fictive), contact (quantitative physical element), repetition and attention (the number of times touch sensed), emotion (emotional charge), sociality (a social phenomenon), and intimacy (the varying degrees of the senses of closeness and distance in social relations). Space manifests itself, phenomenally, as touch. Spatial distance is experienced in something located beyond one's reach. There are social organizations of touch, based on physical and non-physical elements. Place always accompanies touch in materializing caste. Similarly, stench and smell, as experienced in and around waste, open garbage dumps, larger landfills, slums and waste-pickers, has been identified as a main source of ritual pollution, risk, and exclusionary cultural politics. Emergence of dirty, decaying, diseased, and pathogenicity can be addressed by the knowledge of smell.[59]

Dalit movements that charted the interrelationship between caste, nature, and unequal distribution of natural resources marked just a beginning of unraveling complexities of environmental injustices. Notions of environmental human rights – resource and subsistence rights, equal distribution and access to natural resources, and right to information and partnership – have also enthused Dalit movements into articulating environmental issues. Issues of land reform/distribution; caste segregation; discrimination and atrocities in Dalit villages; unequal access to water, forests, commons, and housing; spatial dimensions of Dalit subjugation; and a minimum of environmental health and occupational safety have appeared frequently in Dalit agendas. Dalit livelihood issues and rights in the context of conservation, pollution, extraction of raw materials, alterations of ecosystems, environmental degradation, resource pricing and marketing, impact of climate change, and the making of environmental policy have been prominent in the past two decades. Still, Dalit estrangement goes much deeper and their detachment from mainstream environmental movements continues, in spite of their increased access and participation in the recent past. This means that an understanding is required of the existential, experiential, spatial, and cultural blinders, such as soil, water, air, touch, taste, smell, and space, which carry distributional and participatory software to build Dalit environmental justice and ecology. This also brings to the fore the pluralizing scope and meaning of justice theory vis-à-vis caste issues. In environmental justice, recognition of difference and diversity of experience has integrated some new elements of inequalities, and the processes through which they are reproduced. However, a recognition of caste, and the deep natural, social, and cultural processes involved in the making and unmaking of touch, taste, smell, and senses in a caste society, which crushes people's sense of freedom and belonging and devalues their use, access, and participation in naturescape, can take us towards a richer understanding of environmental justice. Focusing on Dalits as subjects, and the everyday social processes of the construction of their environmental subjectivity, can open up new possibilities of their agency.

Future Archives of Dalit–black Ecologies

There is no document of civilization which is not at the same time a document of barbarism.

– Walter Benjamin

This famous quote suggests that the creation of any civilization involves a heavy cost in terms of lives and ecologies of toiling masses – Dalits and black people – who have been responsible for making that civilization possible. Environmental histories are always an exercise in framing people, place, voices, and visions.[60] The exercise of exploring and forming a future archive of Dalit–black ecologies rests on focusing on their roles in environment, continuing forms of environmental racism and casteism, traditions of environmental justice, preservation of Dalit–black visions of social change, and an updating of arenas, languages, and consequent prescriptions of environmental justice. Such an archive can take inspiration from the exchanges and bonding forged between black and Dalit women's movements. This is a 'margin-to-margin' approach that invites 'different social actors, including scholars and activists, inside a region, nation, or even transnationally to construct shared goals and new bonds of sentiment as well as bodies of knowledge among those most exploited, excluded, or pushed aside'.[61]

In the broader and inclusive framework of environmental injustices, race and caste are widely seen as the key drivers of exclusion and discrimination in social systems and political economies. Several such struggles can be narrated, which mark similarities between African American and Dalit resistances. It has been pointed out that 'access to water is an effective metaphor for characterizing the struggle of the Indian Untouchable and African American to escape oppression, for freedom, justice and equality in the new millennium'.[62] To take a few examples, the Jim Crow laws that lasted officially from 1877 to 1964, and can be felt even in present times, institutionalized the denial of access to public and clean drinking water to African Americans. Martin Luther King Jr's final campaign in 1968 on the striking sanitation workers of Memphis, Tennessee, during which he was assassinated, again shows these linkages. 'Dr. Martin Luther King Jr. and Community on the Move for Equality' called on a 'March for Justice and Jobs' for the city's almost exclusively African American garbage haulers, and demanded equal pay and an end to dirty, hazardous situations. At a time when environmental justice was not on the radar of environment, government, civil rights, public health, or social justice groups, this struggle signified a combination of 'environmental, social and economic justice mission'.[63] This is equally true for Dalit struggles over sanitation and 'polluted' occupations.

Analogies of Dalit–black ecologies can relate to the past and present of castesization and racialization, spatial segregation, sanitation, environmental human rights, organizational mobilization, and movements. Amid this background, three elements give the archive of Dalit–black ecologies its specific character. The first is an attempt to look systematically at the meaning of ecology itself. This entails a group of interrelated questions: How do Dalits and black people figure in ecological studies and amid conditions of existence of living organisms, and what is the interrelationship between these organisms and the environment in which Dalits and black people live? Does an ecological approach provide a critique of the links between centuries of racialization and casteization, as experienced by Dalit–black communities, and the global legitimization of biotic, abiotic, and technogenic changes, created by activities of casteist and racist human societies? How can an examination of ecology as an expanding concept, and its practical application in changing situations over the past century, bear directly on crucial problems of natural and social environment? The second element relates to concerns of dignity and dignified living, rights and justice, in a contemporary criss-cross of environment, caste, race, democracy, participation, recognition, labor, occupation, climate, commons, imaginations, hopes, and struggles in the everyday lives of Dalits and Black people. The third involves a combination of contextual enquiry with concrete steps, with a commitment to environmental rights for every black and Dalit person.

With a substantial number of interventions that explore the nature and meaning of ecologies for Dalits–black people, one central question is on the context in which environmental politics has to be generated. From the perspective of Indian Dalits, it is worth analysing as to how a neoliberal, capitalist growth impacts the natural–social base of lives and livelihoods, and replaces or reproduces structures of hierarchy; how increased mobility, migration, and movement of marginal population enhances or destroys conditions of dignity of labour; and how a global capitalist project legitimizes certain environmental discourses, like green growth and low carbon economy, and de-legitimizes environmental struggles of Dalits on livelihood, occupation, land, water, and forests. Perhaps the most important source feeding into today's Dalit environmental justice movements would be to comprehend the historical trajectories of civil rights movements, anti-caste writings and campaigns, social justice and land rights struggles, labour movements, and the everyday resistances against caste atrocities, segregation, and alienation over natural resources. The archives of black environmental struggles can provide an important reference here.

Dalit–black ecologies can try to capture the positive elements of all these streams, though in the case of Dalits its present dominant themes are developmental, distributional, and life with dignity, recognition, and power. Dalits are some of the most wretched people on earth. They are denied access to basic natural resources – their women spend many hours each day waiting for a faint trickle of dirty water from the polluted municipal tap or contaminated well; Dalit young girls spend their youth scouring the arid and empty landscapes for fuel to cook their single daily meal; landless labourers are transformed into cycle-rickshaw pullers; domestic workers dance to the whims of masters, for whom they represent nothing but hands that perform invisible services; Dalit children are employed in dangerous metal factories; factory workers are locked in during the night shift, and if a fire breaks out scores are trampled to death; people are evicted from their places, and forced to migrate to the pitiless squalor of urban peripheries.[64] Alongside, a Dalit perspective emphasizes a deep urge for a sense of selfhood and freedom as a human being, so that an individual can breathe, inhabit, and cohabit in nature, against natural, physical, spatial, and social structures of domination. These deeper impulses and reckonings mark Dalit environmentalism, which is visible in struggles at local and regional levels. Similar to black environmental struggles, when aggregated together, they voice a Dalit politics of environment justice.

Dalit ecologies is a plural term, as there are multiple ways in which several ex-untouchable castes, haunted by internal divisions and differences, forge larger collectivities, as well as assert individual identities. The plurality of Dalit communities means that such differences are potentially mobilizable in a wide variety of forms, of which issues of natural and physical environment are among the most prominent. Environment is central in the sense that Dalits can feel a fresh sense of life and living, with a different occupation, place, space, politics, and associated changes in their feel, touch, taste, and representation. Environmental right is justice in action, and Dalit environmental agency is a critical context for a new environmental politics. Black environmental struggles have invested their energy in searching for alternative institutions and organizations of governance and power. From organizing local protests to formulating wider agendas, from negotiating on polluting industries and health hazards to legislative reforms, a detailed charter for democratizing governance and power systems has evolved out of these journeys. Black local organizations, publications, performances, and political movements have extensive experimental experiences regarding structures of participation, access, and inclusion in natural and social environments. Such treasures of ecological democracy can be collated to enrich Dalit environmentalism in India.

Notes

1. Dalit Panthers Manifesto (1973), 6–8, https://archive.org/details/idoc_
   pub_dalit-panthers-manifesto/mode/2up?view=theater, accessed on 3
   February 2024.

2. For example, S. D. Kapoor, 'B. R. Ambedkar, W. E. B. Du Bois and the
   Process of Liberation', *Economic and Political Weekly* 38, nos. 51–52 (2003):
   5344–49; Mohan Dass Namishray, *Caste and Race: A Comparative Study
   of B. R. Ambedkar and Martin Luther King* (Jaipur: Rawat Publications,
   2003); Nico Slate, ed., *Black Power Beyond Borders: The Global Dimensions
   of the Black Power Movement* (New York: Palgrave Macmillan, 2012); Suraj
   Yengde and Anand Teltumbde, eds., *The Radical in Ambedkar: Critical
   Reflections* (India: Penguin Random House, 2008).

3. Balmurli Natrajan and Paul Greenough, eds., *Against Stigma: Studies in
   Caste, Race and Justice since Durban* (Hyderabad: Orient Blackswan, 2009).

4. Amongst several publications, some prominent ones are Brian Fair, *Notes
   of a Racial Caste Baby: Color Blindness and the End of Affirmative Action*
   (New York: New York University Press, 1999); Gerald C. Horne, *The End
   of Empires: African Americans and India* (Philadelphia: Temple University
   Press, 2008); Sunita Parikh, *The Politics of Preference: Democratic Institutions
   and Affirmative Action in the United States and India* (Ann Arbor: University
   of Michigan Press, 1997); Vijay Prashad, 'Afro-Dalits of the Earth, Unite!'
   African Studies Review 43, no. 1 (2000): 189–201.

5. Rich literature is available on this issue. For example, Oliver Cromwell
   Cox, *Caste, Class and Race* (New York: Doubleday, 1948); Gunnar Myrdal,
   An Inquiry into the Poverty of Nations (New York: Twentieth Century Fund,
   1968); V. T. Rajashekar, *Dalit: The Black Untouchables of India* (Atlanta:
   Clarity Press, 1995); Peter Robb, ed., *The Concept of Race in South Asia:
   Understanding and Perspectives* (Delhi: Oxford University Press, 1995);
   Kamala Visweswaran, *Un/common Cultures: Racism and the Rearticulation
   of Cultural Difference* (Durham: Duke University Press, 2010); Gyanendra
   Pandey, *A History of Prejudice: Race, Caste, and Difference in India and the
   United States* (Cambridge: Cambridge University Press, 2013); Isabel
   Wilkerson, *Caste: The Origins of Our Discontents* (New York: Random
   House, 2020).

6. Julian Agyeman, Robert D. Bullard, and Bob Evans, eds., *Just Sustainabilities:
   Development in an Unequal World* (Cambridge: MIT Press, 2003), 9.

7. Sunil Menon, 'The Global Dalit, The Indian Black: Cornel West in
   Conversation with Suraj Yengde', *Outlook India*, 3 October 2020, https://
   www.outlookindia.com/magazine/story/india-news-independence-day-

special-whos-indias-george-floyd-heres-exposing-a-racist-india/303552, accessed 12 October 2022.

8. S. Shankar and Charu Gupta, eds., *Caste and Life Narratives* (Delhi: Primus Books, 2009).

9. Patricia Hill Collins, 'It's All in the Family: Intersections of Gender, Race, and Nation', *Hypatia: A Journal of Feminist Philosophy* 13, no. 3 (1998): 63.

10. Christopher W. Wells, ed., *Environmental Justice in Postwar America: A Documentary Reader* (Seattle: University of Washington Press, 2018).

11. Robert D. Bullard, *Dumping in Dixie: Race, Class, and Environmental Quality* (Boulder: Westview Press, 2000).

12. Dorceta E. Taylor, *Race, Class, Gender, and American Environmentalism* (Portland, US Department of Agriculture, 2002).

13. Jedediah Britton-Purdy, 'Environmentalism Was Once a Social-Justice Movement', *The Atlantic*, 7 December 2016, https://www.theatlantic.com/science/archive/2016/12/how-the-environmental-movement-can-recover-its-soul/509831, accessed 3 February 2024.

14. Ronald Sandler and Phaedra C. Pezzullo, eds., *Environmental Justice and Environmentalism: The Social Justice Challenge to the Environmental Movement* (Cambridge: MIT Press, 2007).

15. Sylvia Mayer, ed., *Restoring the Connection to the Natural World: Essays on the African American Environmental Imagination* (Hamburg: LIT VERLAG Munster, 2003), 2.

16. Luke W. Cole and Sheila R. Foster, *From the Ground Up: Environmental Racism and the Rise of the Environmental Justice Movement* (New York: New York University Press, 2001); Filomina Chioma Steady, ed., *Environmental Justice in the New Millennium* (New York: Palgrave Macmillan, 2009).

17. Bullard, *Dumping in Dixie*; Sandler and Pezzullo, *Environmental Justice and Environmentalism*.

18. Andrew Dobson, *Justice and the Environment: Conceptions of Environmental Sustainability and Theories of Distributive Justice* (Oxford: Oxford University Press, 1998); Dorceta E. Taylor, *Toxic Communities: Environmental Racism, Industrial Pollution, and Residential Mobility* (New York: New York University Press, 2014).

19. David Harvey, *Justice, Nature and the Geography of Difference* (Cambridge: Blackwell, 1996).

20. Bruce Braun, 'Towards a New Earth and a New Humanity: Nature, Ontology, Politics', in *David Harvey: A Critical Reader*, ed. Noel Castree and Derek Gregory (Oxford: Blackwell, 2006), 191–222.

21. William Cronon, ed., *Uncommon Ground: Rethinking the Human Place in Nature* (New York: W. W. Norton & Company, 1996).

22. Rob White, *Environmental Harm: An Eco-justice Perspective* (Bristol: Polity Press, 2013).

23. Francis O. Adeola, 'Cross-national Environmental Injustice and Human Rights Issues: A Review of Evidence in the Developing World', *American Behavioral Scientist* 43, no. 4 (2000): 696–706; John Byrne, Leigh Glover, and Cecilia Martinez, *Environmental Justice: Discourses in International Political Economy* (New Brunswick, NJ: Transaction, 2002).

24. Joan Martinez-Alier, Leah Temper, Daniela Del Bene, and Arnim Scheidel, 'Is There a Global Environmental Justice Movement?' *Journal of Peasant Studies* 43, no. 3 (2016): 731–55.

25. Bunyan Bryant, ed., *Environmental Justice: Issues, Policies, and Solutions* (Washington, DC: Island Press, 1995).

26. There are a large number of studies in this area. I refer to only a few: Chris O. Ikporukpo, 'Petroleum, Fiscal Federalism and Environmental Justice in Nigeria', *Space and Polity* 8, no. 3 (2004): 321–54; Mei-Fang Fan, 'Environmental Justice and Nuclear Waste Conflicts in Taiwan', *Environmental Politics* 15, no. 3 (2006): 417–34; Ilan Alleson and Stuart Schoenfeld, 'Environmental Justice and Peacebuilding in the Middle East', *Peace Review* 19, no. 3 (2007): 371–79; Sara E. Grineski and Timothy W. Collins, 'Exploring Patterns of Environmental Injustice in the Global South: Maquiladoras in Ciudad Juarez, Mexico', *Population and Environment* 29, no. 6 (2008): 247–70.

27. Carl Anthony, 'The Environmental Justice Movement: An Activist's Perspective', in *Power, Justice, and the Environment*, ed. David N. Pellow and Robert J. Brulle (Cambridge: MIT Press, 2005), 92.

28. S. Ravi Rajan, 'Environmental Justice in India', *Environmental Justice* 7, no. 5 (2014): 120.

29. Ramachandra Guha and Juan Martinez-Alier, *Varieties of Environmentalism: Essays North and South* (Delhi: Oxford University Press, 1998).

30. Government of Madhya Pradesh, *The Bhopal Document: Creating a New Course for Dalits for the 21st Century* (Bhopal: Government of Madhya Pradesh, 2002).

31. M. Bhimraj, 'A Dalit Critique of Environmental Justice in India', in *Contemporary Environmental Concerns* (Punjab: Rajiv Gandhi National University of Law, 2020), 89–120.

32. M. S. Vani, Rohit Asthana, Narayan Belbase, L. B. Thapa, Patricia Moore, and Firuza Pastakia, *Environmental Justice and Rural Communities: Studies from India and Nepal* (Bangkok: IUCN, 2007).

33. Quoted in Narendra Jadhav, ed., *Ambedkar Speaks: 301 Seminal Speeches* (New Delhi: Konark Publications, 2013).

34. Anupama Rao, *The Caste Question: Dalits and the Politics of Modern India* (Ranikhet: Permanent Black, 2009).

35. For a deeper understanding of caste and water, see Mukul Sharma, *Caste and Nature: Dalits and Indian Environmental Politics* (Delhi: Oxford University Press, 2017), 161–211.

36. Badri Narayan, *The Making of the Dalit Public in North India: Uttar Pradesh, 1950–Present* (New Delhi: Oxford University Press, 2011).

37. Sanitation Workers Movement (In Hindi), 'Bhim March Daily Reports (In Hindi)', 2016, https://www.safaikarmachariandolan.org/resources, accessed on 3 February 2024.

38. Bezwada Wilson and Bhasha Singh, *The Long March to Eliminate Manual Scavenging: India Exclusion Report 2016* (Delhi: Yoda Press, 2016), 298–319.

39. Rashtriya Garima Abhiyan (National Dignity Campaign), 'Women Liberation March' (in Hindi), 2013.

40. K. B. Saxena, *Prevention of Atrocities against Scheduled Castes: Policy and Performance* (Delhi: National Human Rights Commission, 2004).

41. Kenneth Bo Nielsen, Siddharth Sareen, and Patrik Oskarsson, 'The Politics of Caste in India's New Land Wars', *Journal of Contemporary Asia* 50, no. 5 (2020): 684–95.

42. For details, Ravi Adagale, 'Water and Violence against Dalits in Maharashtra: A Multi-case Approach', *Social Change* 50, no. 3 (2020): 399–415; Thomas Crowley, 'Leisure, Festival, Revolution: Ambedkarite Productions of Space', *Caste: A Global Journal on Social Exclusion* 1, no. 2 (2020): 31–50.

43. For details, see Goldy M. George, *Caste Discrimination and Dalit Rights over Natural Resources* (Chhattisgarh: Dalit Liberation Front, 2011 in Hindi); Anand Vaidya, 'This Forests Belong to Us' (in Hindi)', *Seminar* (2017), https://www.india-seminar.com/2017/690/690_anand_vaidya.htm, accessed 3 February 2024.

44. Ravikumar, *Venomous Touch: Notes on Caste, Culture and Politics* (Kolkata: Samya, 2009), 246.

45. D. R. Nagraj, *The Flaming Feet and Other Essays: The Dalit Movement in India* (Ranikhet: Permanent Black, 2010), 133.

46. T. H. P. Chentharassery, 'Struggle of Freedom', in *No Alphabet in Sight: New Dalit Writing from South India*, ed. K. Satyanarayana and Susie Tharu (New Delhi: Penguin Books, 2011), 385–92.

47. Gyanendra Pandey, 'Politics of Difference: Reflections on Dalit and African American Struggles', *Economic and Political Weekly* 45, no. 19 (2010): 66.

48. Jesus F. Chairez-Garza, Mabel Denzin Gergan, Malini Ranganathan, and Pavithra Vasudevan, 'Introduction to the Special Issue: Rethinking

Difference in India Through Racialization', *Ethnic and Racial Studies* 45, no. 2 (2022): 193.

49. Mukul Sharma and Ashok Bharti, 'In Lieu of an Introduction', in *Defining Dignity: An Anthology of Dreams, Hopes and Struggles*, ed. Mukul Sharma and Sana Das (Delhi: World Dignity Forum, 2005), 2–12.

50. Lorraine Daston, 'The Naturalized Female Intellect', *Science in Context* 5, no. 2 (1992): 209.

51. Gopal Guru and Sundar Sarukkai, *Experience, Caste, and the Everyday Social* (New Delhi: Oxford University Press, 2019).

52. Timothy V. Kaufman-Osborn, 'Teasing Feminist Sense from Experience', *Hypatia* 8, no. 2 (1993): 124–44.

53. Kim Allen, Vinci Daro, and Dorothy C. Holland, 'Becoming an Environmental Justice Activist', in *Environmental Justice and Environmentalism*, ed. Ronald Sandler and Phaedra C. Pezzullo (Cambridge, MA: MIT Press, 2007), 127.

54. Sundar Sarukkai, 'Experience and Theory: From Habermas to Gopal Guru', in *The Cracked Mirror: An Indian Debate on Experience and Theory*, ed. Gopal Guru and Sundar Sarukkai (New Delhi: Oxford University Press, 2012), 36.

55. Gopal Guru, 'Experience, Space, and Justice', in *The Cracked Mirror*, ed. Guru and Sarukkai (New Delhi: Oxford University Press, 2012), 71–106.

56. Nagraj, *The Flaming Feet and Other Essays*, 162.

57. Malini Ranganathan, 'Caste, Racialization, and the Making of Environmental Unfreedoms in Urban India', *Ethnic and Racial Studies* 45, no. 2 (2022): 257–77.

58. Aniket Jaaware, *Practicing Caste: On Touching and Not Touching* (New York: Fordham University Press, 2019).

59. Assa Doron, 'Stench and Sensibilities: On Living with Waste, Animals and Microbes in India', *Australian Journal of Anthropology* 32, no. S1 (2021): 23–41.

60. David Schlosberg, 'The Justice of Environmental Justice: Reconciling Equity, Recognition, and Participation in a Political Movement', in *Moral and Political Reasoning in Environmental Practice*, ed. Andrew Light and Avner De-Shalit (Cambridge, MA: MIT Press, 2003), 77–106.

61. Shailaja Paik, 'Building Bridges: Articulating Dalit and African American Women's Solidarity', *WSQ: Women's Studies Quarterly* 42, nos. 3–4 (2014): 75–76.

62. There are many publications on this. For example, Ronald E. Hall and Neha Mishra, 'Ambedkar and King: The Subjugation of Caste or Race vis-à-vis Colourism', in *The Radical in Ambedkar*, ed. Suraj Yengde and Anand Teltumbde (Gurgaon: Penguin Random House, 2018); Poppy

Noor, 'Detroit Families Still Without Clean Water Despite Shutoffs Being Lifted'", *The Guardian*, 20 May 2020, https://bit.ly/34GJPQI, accessed 16 May 2022; Susan Spronk, 'Covid-19 and Structural Inequalities: Class, Gender, Race and Water Justice', in *Public Water and Covid-19: Dark Clouds and Silver Linings*, ed. David A. McDonald, Susan J. Spronk, and Daniel Chavez (Kingston: Municipal Services Project, 2020), 25-48.

63. Glenn S. Johnson, 'Environmental Justice: A Brief History and Overview', in *Environmental Justice in the New Millennium*, ed. Filomina Chioma Steady (New York: Palgrave Macmillan, 2009), 17.

64. Jeremy Seabrook, *Class, Caste and Hierarchies* (Jaipur: Rawat Publications, 2005).

Bibliography

Adagale, Ravi. 'Water and Violence against Dalits in Maharashtra: A Multi-Case Approach'. *Social Change* 50, no. 3 (2020): 399–415.

Adeola, Francis O. 'Cross-national Environmental Injustice and Human Rights Issues: A Review of Evidence in the Developing World'. *American Behavioral Scientist* 43, no. 4 (2000): 696–706.

Agarwal, Anil and Ajit Chak, eds. *State of India's Environment, A Citizen's Report: Floods, Flood Plains and Environmental Myths*. Delhi: Center for Science and Environment, 1991.

Agyeman, Julian, Robert D. Bullard, and Bob Evans, eds. *Just Sustainabilities: Development in an Unequal World*. Cambridge, MA: MIT Press, 2003.

Ahmad, Zarin. *Delhi's Meatscapes: Muslim Butchers in a Transforming Mega-City*. New Delhi: Oxford University Press, 2018.

Alam, Mahmood. 'Problems and Prospects of Leather Industry in U.P.' A thesis submitted for Master of Philosophy, Aligarh Muslim University, Aligarh, India, 1991.

Ali, Syed Asif. 'The Urban Geography of Kanpur'. A thesis submitted for Master of Philosophy, University of London, July 1970.

Allen, Kim, Vinci Daro, and Dorothy C. Holland. 'Becoming an Environmental Justice Activist'. In *Environmental Justice and Environmentalism*, edited by Ronald Sandler and Phaedra C. Pezzullo, 105–34. Cambridge, MA: MIT Press, 2007.

Alleson, Ilan and Stuart Schoenfeld. 'Environmental Justice and Peacebuilding in the Middle East'. *Peace Review* 19, no. 3 (2007): 371–79.

Alter, Stephen, ed. *Writing Outdoors: A Natural Reader*. New Delhi: WWF-India, 2014.

Alvares, Claude. *Decolonising History: Technology and Culture in India, China and the West 1492 to the Present Day*. Goa: The Other India Press, 1991.

Ambedkar, B. R. 'Annihilation of Caste'. In *The Essential Writings of B. R. Ambedkar*, edited by Valerian Rodrigues, 263–305. New Delhi: Oxford University Press, 2002.

———. 'Philosophy of Hinduism'. In *Dr Babasaheb Ambedkar Writings and Speeches*, edited by Vasant Moon, vol. 3, 3–92. Mumbai: Government of Maharashtra, 1987.

Anand, S., ed. *Touchable Tales: Publishing and Reading Dalit Literature*. Chennai: Navayana, 2003.

Anderson, B. D. 'Concrete Kingdom: Sculptures by Nek Chand'. *Folk Art* 31, nos. 1–2 (2006): 42–49.

Anthony, Carl. 'Reflections on the Purposes and Meanings of African American Environmental History'. In *'To Love the Wind and the Rain': African Americans and Environmental History*, edited by Dianne D. Glave and Mark Stoll, 200–09. Pittsburgh: University of Pittsburgh Press, 2006.

———. 'The Environmental Justice Movement: An Activist's Perspective'. In *Power, Justice, and the Environment*, edited by David N. Pellow and Robert J. Brulle, 91–100. Cambridge, MA: MIT Press, 2005.

Appadurai, Arjun, Frank J. Korom, and Margaret A. Mills. 'Introduction'. In *Gender, Genre, and Power in South Asian Expressive Traditions*, edited by Arjun Appadurai, Frank. J. Korom, and Margaret A. Mills, 3–29. Philadelphia: University of Pennsylvania Press, 1991.

Arnold, David. *Everyday Technology: Machines and the Making of India's Modernity*. Chicago: The University of Chicago Press, 2013.

Asher, Kiran. *Black and Green: Afro-Colombians, Development, and Nature in the Pacific Lowlands*. Durham: Duke University Press, 2009.

Aulakh, M. *The Rock Garden*. Hyderabad: Tagore Publishers, 1986.

Bakhtin, Mikhail. *The Dialogic Imagination*. Translated by Caryl Emerson and Michael Holquist. Austin: University of Texas Press, 1981.

Balasubramanian, Muniyandi. 'Climate Change, Famine, and Low-Income Communities Challenge Sustainable Development Goals'. *The Lancet* 2, no. 10 (October 2018): E421–E422.

Balasubramanian, M., Manjunatha Munishamappa, Remadevi O. K., Vinayakumar K. Hombalegowda, and Rittu Kakkar. *Climate Change and Its Impact on Vulnerable Communities: A Case Study of Karnataka*. Bangalore: Institute for Social and Economic Change, 2019.

Bama. *Karukku*. Translated from Tamil by Lakshmi Holmstrom. New Delhi: Oxford University Press, 2012.

Bandyopadhyay, Sekhar. 'Partition and the Ruptures in Dalit Identity Politics in Bengal'. *Asian Studies Review* 33, no. 4 (2009): 455–67.

Bandyopadhyay, Soumyen and Iain Jackson. *The Collection, the Ruin and the Theatre: Architecture, Sculpture and Landscape in Nek Chand's Rock Garden*. Liverpool: Liverpool University Press, 2007.

Banks, Adam J. *Race, Rhetoric, and Technology: Searching for Higher Ground*. New Jersey: LEA and NCTE, 2006.

Bardhan, Kalpana. Introduction and Afterword to *A River Called Titash*, edited by Adwaita Mallabarman, 1–4, 259–62. Berkeley: University of California Press, 1993.

Bargi, Drishadwati. 'Understanding "Dalit Chetna" in Adwaita Mallabarman's *Titash Ekti Nadir Naam*, A River Called Titash'. *Contemporary Voice of Dalit* 8, no. 1 (2016): 90–104.

Bauman, Richard. *Verbal Art as Performance*. Rowley: Newbury House, 1977.

Baviskar, Amita, ed. *Waterlines: The Penguin Book of River Writings*. New Delhi: Penguin Books, 2003.

Beck, Brenda E. F. *The Three Twins: The Telling of a South-Indian Folk Epic*. Bloomington: Indiana University Press, 1982.

Bennett, Jane. *Vibrant Matter: A Political Ecology of Things*. Durham: Duke University Press, 2010.

Bennett, Lynn, Dilli Ram Dahal, and Pav Govindasamy. *Caste, Ethnic and Regional Identity in Nepal: Further Analysis of the 2006 Nepal Demographic and Health Survey*. Maryland: Macro International Inc., 2008.

Berry, Thomas. *The Dream of the Earth*. Oakland: Sierra Club Books, 2016.

Bhagavan, Manu and Anne Feldhaus. *Claiming Power from Below: Dalits and the Subaltern Question in India*. New Delhi: Oxford University Press, 2008.

Bhan, Gautam. *In the Public's Interest: Eviction, Citizenship, and Inequality in Contemporary Delhi*. Athens: The University of Georgia Press, 2006.

Bharne, Vinayak. *The Emerging Asian City: Concomitant Urbanities and Urbanisms*. New York: Routledge, 2013.

Bharucha, Rustom. *Rajasthan, an Oral History: Conversations with Komal Kothari*. Delhi: Penguin, 2003.

Bhatia, Nandi. *Acts of Authority/Acts of Resistance: Theatre and Politics in Colonial and Postcolonial India*. Ann Arbor: The University of Michigan Press, 2004.

Bhattacharya, Shahana. 'Transforming Skin, Changing Caste: Technical Education in Leather Production in India, 1900–1950'. *Indian Economic and Social History Review* 55, no. 3 (2018): 307–43.

Bhatti, S. S. *Rock Garden in Chandigarh: A Critical Evaluation of the Work of Nek Chand*. Chandigarh: White Falcon Publishing, 2018.

Bhimraj, M. 'A Dalit Critique of Environmental Justice in India'. In *Contemporary Environmental Concerns*, 89–120. Punjab: Rajiv Gandhi National University of Law, 2020.

Bidwai, Praful. *The Politics of Climate Change and the Global Crisis: Mortgaging Our Futures*. New Delhi: Orient BlackSwan, 2012.

Bindra, Prerna Singh. *The Vanishing: India's Wildlife Crisis*. New Delhi: Penguin Books, 2017.

Biswas, Manohar Mouli. *Surviving in My World: Growing Up Dalit in Bengal*. Translated from Bengali by Angana Dutta and Jaydeep Sarangi. Kolkata: Samya, 2011.

Biswas, Sravani. 'Nature and Humans in the Imagination of Bengali Intellectuals of 1930s–50s'. *Studies on Asia* 1, no. 2 (2011): 15–34.

Blackburn, Stuart H. *Himalayan Tribal Tales: Oral Tradition and Culture in the Apatani Valley*. Leiden: BRILL, 2008.

———. *Singing of Birth and Death: Texts in Performance*. Philadelphia: University of Pennsylvania Press, 1988.

———. 'The Folk Hero and Class Interests in Tamil Heroic Ballads'. *Asian Folklore Studies* 37, no. 1 (1978): 131–49.

Blackburn, Stuart H. and Joyce B. Flueckiger. 'Introduction'. In *Oral Epics in India*, edited by Stuart H. Blackburn, Peter J. Claus, Joyce B. Flueckiger, and Susan S. Wadley, 1–11. Berkeley: University of California Press, 1989.

Bond, Ruskin. *The Book of Nature*. New Delhi: Penguin India, 2008.

Bonfitto, Tracy Ann Buck. *The Rock Garden: A Study of Memory, Place-Making, and Community in Chandigarh, India*. Los Angeles: University of California, 2018. Accessed on 4 February 2024. https://escholarship.org/uc/item/80j5t8x4.

Bora, Ram Singh. 'Migrant Informal Workers: A Study of Delhi and Satellite Towns'. *Modern Economy* 5, no. 5 (2014): 562–79.

Bottomore, Tom. *A Dictionary of Marxist Thought*. New Delhi: Oxford University Press, 1983.

Boyle, Godfrey. *Community Technology*. Milton Keynes: Open University Press, 1978.

Braun, Bruce. 'Towards a New Earth and a New Humanity: Nature, Ontology, Politics'. In *David Harvey: A Critical Reader*, edited by Noel Castree and Derek Gregory, 199–222. Oxford: Blackwell, 2006.

Bridges, Kelly. 'Climate Change, Scheduled Caste, and Scheduled Tribes: Analyzing Socioeconomic and Climate Change Vulnerabilities Amongst Female Farmers in Rural Madhya Pradesh'. *In Situ* no. 1 (May 2016).

Britton-Purdy, Jedediah. 'Environmentalism Was Once a Social-Justice Movement'. *The Atlantic*, 7 December 2016. Accessed on 3 February 2024. https://www.theatlantic.com/science/archive/2016/12/how-the-environmental-movement-can-recover-its-soul/509831.

Brooke. *Brick by Brick: Environment, Human Labour and Animal Welfare*. London: Brooke, 2017.

———. *The Danger of Heat Stress in India's Brick Kilns*. London: Brooke Action for Working Horses and Donkeys, 2019. Accessed on 4 February 2024. https://www.thebrooke.org/our-work/india/danger-heat-stress-indias-brick-kilns.

Bryant, Bunyan, ed. *Environmental Justice: Issues, Policies, and Solutions*. Washington, DC: Island Press, 1995.

Buell, Lawrence. *The Future of Environmental Criticism: Environmental Crisis and Literary Imagination*. New Jersey: Blackwell, 2005.

Bullard, Robert D. *Dumping in Dixie: Race, Class, and Environmental Quality*. Boulder, CO: Westview Press, 2000.

———. *The Quest for Environmental Justice: Human Rights and the Politics of Pollution*. San Francisco: Sierra Club Books, 2005.

Bullard, Robert D. and Beverly Wright, eds. *Race, Place and Environmental Justice after Hurricane Katrina*. Boulder: Westview Press, 2009.

Byapari, Manoranjan Byapari. *Interrogating My Chandal Life: An Autobiography of a Dalit*. Translated from Bengali by Sipra Mukherjee. New Delhi: Sage, 2018.

Byrne, John, Leigh Glover, and Cecilia Martinez. *Environmental Justice: Discourses in International Political Economy*. New Brunswick, NJ: Transaction, 2002.

Carruthers, David V., ed. *Environmental Justice in Latin America: Problems, Promise, and Practice*. Cambridge: MIT Press, 2008.

Census of India 2011. New Delhi: Government of India, 2011. Accessed on 8 February 2020. www.censusinida.gov.in.

Chairez-Garza, Jesus F., Mabel Denzin Gergan, Malini Ranganathan, and Pavithra Vasudevan. 'Introduction to the Special Issue: Rethinking Difference in India through Racialization'. *Ethnic and Racial Studies* 45, no. 2 (2022): 193–215.

Chakrabarty, Dipesh. *The Climate of History in a Planetary Age*. Chicago: The University of Chicago Press, 2021.

———. 'The Dalit Body: A Reading for the Anthropocene'. In *The Empire of Disgust: Prejudice, Discrimination, and Policy in India and the US*, edited by Zoya Hasan, Aziz Z. Huq, Martha C. Nussbaum, and Vidhu Verma, 1–20. Delhi: Oxford University Press, 2018.

Chakravarty, Surajit and Rohit Negi, eds. *Space, Planning and Everyday Contestations in Delhi*. Delhi: Springer India, 2016.

Chalana, Manish. 'Chandigarh: City and Periphery'. *Journal of Planning History* 14, no. 1 (2015): 62–84.

Chentharassery, T. H. P. 'Struggle of Freedom'. In *No Alphabet in Sight: New Dalit Writing from South India*, edited by K. Satyanarayana and Susie Tharu, 385–92. New Delhi: Penguin Books, 2011.

Chintan. *Cooling Agents: An Examination of the Role of the Informal Recycling Sector in Mitigating Climate Change*. New Delhi: Safai Sena, The Advocacy Project, Chintan, 2009.

———. *Wastepickers: Delhi's Forgotten Environmentalists?* New Delhi: Chintan, 2019.

Chopra, Suneet. 'Bondage in a Green Revolution Area: A Study of Muzaffarnagar Brick-Kiln Workers'. *Social Scientist* 10, no. 3 (1982): 38–55.

Clare, Stephanie. 'Geopower: The Politics of Life and Land in Frantz Fanon's Writing'. *Diacritics* 41, no. 4 (2013).

Climate Change Programme. *Climate Change and Agriculture in India*. New Delhi: Ministry of Science and Technology, Government of India, 2016.

Cock, Jacklyn and David Fig. 'The Impact of Globalisation on Environmental Politics in South Africa, 1990–2002'. *African Sociological Review* 5, no. 2 (2021): 15–35.

Cole, Luke W. and Sheila R. Foster. *From the Ground Up: Environmental Racism and the Rise of the Environmental Justice Movement*. New York: New York University Press, 2001.

Collins, Patricia Hill. 'It's All in the Family: Intersections of Gender, Race, and Nation'. *Hypatia: A Journal of Feminist Philosophy* 13, no. 3 (1998): 62–82.

Colopy, Cheryl. *Dirty, Sacred Rivers: Confronting South Asia's Water Crisis*. New Delhi: Oxford University Press, 2012.

Corbusier, Le. 'Chand Indes Urb Plan'. General No. 11/30887. Paris: Foundation Le Corbusier, 1953.

———. *The City of To-Morrow and Its Planning*. Cambridge: MIT Press, 1971.

Cortesi, Luisa. 'The Muddy Semiotics of Mud'. *Journal of Political Ecology* 25, no. 1 (November 2018): 617–37.

Cox, Oliver Cromwell. *Caste, Class and Race*. New York: Doubleday, 1948.

Cronon, William, ed. *Uncommon Ground: Rethinking the Human Place in Nature*. New York: W. W. Norton & Company, 1996.

———, ed. *Uncommon Ground: Towards Reinventing Nature*. New York: W. W. Norton & Company, 1995.

Crooke, William. *The Tribes and Castes of the North-Western Provinces and Oudh*. Vol. 3. Calcutta: Office of the Superintendent of Government Printing, 1896.

Crowley, Thomas. 'Leisure, Festival, Revolution: Ambedkarite Productions of Space'. *Caste: A Global Journal on Social Exclusion* 1, no. 2 (2020): 31–50.

Dalit Panthers Manifesto. 1973. Accessed on 4 February 2024. https://archive.org/details/idoc_pub_dalit-panthers-manifesto/mode/2up?view=theater.

Damodaran, Harish. *India's New Capitalists: Caste, Business, and Industry in a Modern Nation*. New York: Palgrave Macmillan, 2008.

Damodaran, Sumangala and Pallavi Mansingh. *Leather Industry in India*. Delhi: Centre for Education and Communication, 2008.

Daston, Lorraine. 'The Naturalized Female Intellect'. *Science in Context* 5, no. 2 (1992): 209–35.

Davis, Coralynn V. 'Pond-Women Revelations: The Subaltern Registers in Maithil Women's Expressive Forms'. *Journal of American Folklore* 121, no. 481 (2008): 286–318.

Davis, Heather and Etienne Turpin. 'Art and Death: Lives between the Fifth Assessment and the Sixth Extinction'. In *Art in the Anthropocene: Encounters among Aesthetics, Politics, Environments and Epistemologies*, edited by Heather Davis and Etienne Turpin, 3–29. London: Open Humanities Press, 2015.

Delhi Master Plans. Accessed on 4 February 2024. https://dda.gov.in/about-master-plan.

DeLoughrey, Elizabeth M. *Allegories of Anthropocene*. Durham: Duke University Press, 2019.

Deming, Alison H. and Lauret E. Savoy. *Colors of Nature: Culture, Identity, and the Natural World*. Minnesota: Milkweed, 2011.

Deshpande, Ashwini. *The Grammar of Caste: Economic Discrimination in Contemporary India*. New Delhi: Oxford University Press, 2011.

Detraz, Nicole. 'Threats or Vulnerabilities? Assessing the Link between Climate Change and Security'. *Global Environmental Politics* 11, no. 3 (2011): 104–20.

Dharmalingam, A. 'Conditions of Brick Workers in a South Indian Village'. *Economic and Political Weekly* 30, no. 47 (1995): 3014–18.

Dharwadker, Vinay, ed. *The Collected Essays of A. K. Ramanujan*. New Delhi: Oxford University Press, 2006.

Dhasal, Namdeo. *A Current of Blood*. New Delhi: Navayana, 2007.

Dickson, David. *Alternative Technology and the Politics of Technical Change*. London: Fontana, 1975.

Directorate of Census Operations. *District Census Handbook Jhajjar: Village and Town Wise Primary Census Abstract (PCA)*. Chandigarh: Government of India, 2011.

Dobson, Andrew. *Justice and the Environment: Conceptions of Environmental Sustainability and Theories of Distributive Justice*. Oxford: Oxford University Press, 1998.

Doron, Assa. 'Stench and Sensibilities: On Living with Waste, Animals and Microbes in India'. *Australian Journal of Anthropology* 32, no. S1 (February 2021): 23–41.

Doron, Assa and Robin Jeffrey. *Waste of a Nation: Garbage and Growth in India*. Cambridge: Harvard University Press, 2018.

Douglas, Mary. *Purity and Danger: An Analysis of the Concepts of Pollution and Taboo.* New York: Routledge, 1966.

Drèze, Jean, Meera Samson, and Satyajit Singh. *The Dam and the Nation: Displacement and Resettlement in the Narmada Valley.* Delhi: Oxford University Press, 1997.

Dubash, Navroz K., ed. *Handbook of Climate Change and India: Development, Politics and Governance.* Oxon: Earthscan, 2012.

———, ed. *India in a Warming World: Integrating Climate Change and Development.* New Delhi: Oxford University Press, 2019.

Dundes, Alan. *The Study of Folklore.* London: Prentice-Hall International, INC, 1965.

Eck, Diana L. *India: A Sacred Geography.* New York: Three Rivers Press, 2012.

Edquist, Charles. *Capitalism, Socialism and Technology.* London: Zed Books, 1985.

Eil, Andrew, Jie Li, Prajwal Baral, and Eri Saikawa. *Dirty Stacks, High Stakes: An Overview of Brick Sector in South Asia.* Washington, DC: The World Bank, 2020.

Enloe, Cynthia. *Seriously! Investigating Crashes and Crises as If Women Mattered.* Berkeley: University of California Press, 2013.

Fair, Brian. *Notes of a Racial Caste Baby: Color Blindness and the End of Affirmative Action.* New York: New York University Press, 1999.

Fan, Mei-Fang. 'Environmental Justice and Nuclear Waste Conflicts in Taiwan'. *Environmental Politics* 15, no. 3 (2006): 417–34.

Feldhaus, Anne, Ramdas Atkar, and Rajaram Zagade, eds. and trans. *Say to the Sun, 'Don't Rise,' and to the Moon, 'Don't Set': Two Oral Narratives from the Countryside of Maharashtra.* New York: Oxford University Press, 2014.

Fernandes, Walter. *Sixty Years of Development Induced Displacement in India: Impacts and the Search for Alternatives.* Delhi: Oxford University Press, 2008.

Fernandez, Marilyn. *The New Frontier: Merit vs Caste in Indian IT Sector.* Delhi: Oxford University Press, 2018.

Figueiredo, Marina Dantas de, Fábio Freitas Schilling Marquesan, and José Miguel Imas. 'Anthropocene and "Development": Intertwined Trajectories since the Beginning of the Great Acceleration'. *Revista de Administração Contemporânea – RAC* 24, no. 5/2 (2020): 400–13.

Figueroa, Robert M. 'Bivalent Environmental Justice and the Culture of Poverty'. *Rutgers University Journal of Law and Urban Policy* 1, no. 1 (2003): 27–44.

Flueckiger, Joyce Burkhalter. *Gender and Genre in the Folklore of Middle India.* Itacha: Cornell University Press, 1996.

Fraser, Nancy. *Justice Interruptus: Critical Reflections on the 'Postsocialist' Condition.* New York: Routledge, 1997.

Fry, Maxwell. 'Chandigarh–New Capital City'. *Architectural Record* (June 1955), 143.

Gandhi, Tara, ed. *A Bird's Eye View: Collected Essays and Shorter Writings of Salim Ali*. Vols. 1 and 2. Delhi: Permanent Black, 2006 and 2007.

Ganguly, Debjani. 'Pain, Personhood and the Collective: Dalit Life Narratives'. *Asian Studies Review* 33, no. 4 (December 2009): 429–42.

Geetha, V. and S. V. Rajadurai. *Towards a Non-Brahmin Millennium: From Iyothee Thass to Periyar*. Kolkata: Samya, 2008.

George, Goldy M. *Caste Discrimination and Dalit Rights over Natural Resources*. Chhattisgarh: Dalit Liberation Front (in Hindi), 2011.

Ghai, Rahul, Arvind Kumar Mishra, and Sanjay Kumar. *The Marginalized Self: Tales of Resistance of a Community*, edited by Rahul Ghai, Arvind K. Mishra and Sanjay Kumar. Delhi: Primus Books, 2020.

Ghosh, Amitav. *The Great Derangement: Climate Change and the Unthinkable*. Chicago: Chicago University Press, 2016.

Glave, Dianne D. *Rooted in the Earth: Reclaiming the African American Environmental Heritage*. Chicago, Illinois: Lawrence Hill Books, 2010.

———. 'Rural African American Women, Gardening, and Progressive Reform in the South'. In *'To Love the Wind and the Rain': African Americans and Environmental History*, edited by Dianne D. Glave and Mark Stoll, 37–50. Pittsburgh: University of Pittsburgh Press, 2006.

Glotfelty, Cheryll. 'Introduction'. In *The Ecocriticism Reader: Landmarks in Literary Ecology*, edited by Cheryll Glotfelty and Harold Fromm, xv–xxxvii. Athens: University of Georgia Press, 1996.

Gooptu, Nandini. *The Politics of the Urban Poor in Early Twentieth-Century India*. Cambridge: Cambridge University Press, 2001.

Gould, Harold A. 'The Adaptive Functions of Caste in Contemporary Indian Society'. *Asian Survey* 3, no. 9 (1963): 427–438.

Government of Haryana. *Haryana State Action Plan on Climate Change*. Chandigarh, 2011.

Government of Haryana, Department of Revenue and Disaster Management. *District Disaster Management Plan Jhajjar*. Jhajjar: Haryana Institute of Public Management and District Administration, 2018.

Government of India. *Shramshakti: Report of the National Commission on Self-Employed Women and Women in the Informal Sector*. New Delhi, 1988.

———. *The National Action Plan on Climate Change*. Accessed on 3 February 2024. https://envt.kerala.gov.in/wp-content/uploads/2019/05/National-Action-Plan-on-Climate-Change.pdf.

Government of India, Ministry of MSME. *Brief Industrial Profile of Jhajjar District*. Karnal: MSME-Development Institute, n.d.

Government of Madhya Pradesh. *The Bhopal Document: Changing a New Course for Dalits for the 21st Century*. Bhopal: Government of Madhya Pradesh, 2002.

Gramsci, Antonio. *Selections from the Prison Notebooks*. London: Lawrence and Wishart, 1971.

Grierson, George Abraham. 'Selected Specimens of the Bihārī Language'. *Zeitschrift der Deutschen Morgenländischen Gesellschaft* 39, no. 4 (1885): 617–73.

Grineski, Sara E. and Timothy W. Collins. 'Exploring Patterns of Environmental Injustice in the Global South: Maquiladoras in Ciudad Juarez, Mexico'. *Population and Environment* 29, no. 6 (2008): 247–70.

Guha, Ramachandra, ed. *Nature's Spokesman: M. Krishnan and Indian Wildlife*. New Delhi: Oxford University Press, 1997.

———. *Savaging the Civilized: Verrier Elwin, His Tribals, and India*. UK: Penguin Books, 2014.

Guha, Ramachandra and Juan Martinez-Alier. *Varieties of Environmentalism: Essays North and South*. Delhi: Oxford University Press, 1998.

Gulati, Leela. 'Female Labour in the Unorganised Sector: Profile of a Brick Worker'. *Economic and Political Weekly* 14, no. 16 (1979): 744–52.

Gulati, Leela and Mitu Gulati. 'Female Labour in the Unorganised Sector: The Brick Worker Revisited'. *Economic and Political Weekly* 32, no. 18 (1997): 968–71.

Gupta, Charu. *Sexuality, Obscenity, Community: Women, Muslims, and the Hindu Public in Colonial India*. Delhi: Permanent Black, 2001.

———. *The Gender of Caste: Representing Dalits in Print*. Ranikhet: Permanent Black, 2016.

Gupta, Ishika, Prakashan Chellattan Veettil, and Stijn Speelman. 'Caste, Social Networks and Variety Adoption'. *Journal of South Asian Development* 15, no. 2 (2020): 155–83.

Gupta, Shivam, Rocky Gupta, and Ronak Tamra. *Challenges Faced by Leather Industry in Kanpur*. Kanpur: Indian Institute of Technology, 2007.

Guru, Gopal. 'Archaeology of Untouchability'. *Economic and Political Weekly* 44, no. 27 (2009): 49–56.

———. 'Experience, Space, and Justice'. In *The Cracked Mirror: An Indian Debate on Experience and Theory*, edited by Gopal Guru and Sundar Sarukkai, 71–106. New Delhi: Oxford University Press, 2012.

———. 'Moral Significance of Justice: Foregrounding Environmentalism in India'. Public Lecture at Nehru Memorial Museum and Library, New Delhi, 26 September 2013.

Guru, Gopal and Sundar Sarukkai. *Experience, Caste, and the Everyday Social*. New Delhi: Oxford University Press, 2019.

Haberman, David L. 'River of Love in an Age of Pollution'. In *Hinduism and Ecology: The Intersection of Earth, Sky, and Water*, edited by Christopher Key Chapple and Mary Evelyn Tucker, 339–54. New Delhi: Oxford University Press, 2000.

Hall, Ronald E. and Neha Mishra. 'Ambedkar and King: The Subjugation of Caste or Race vis-à-vis Colourism'. In *The Radical in Ambedkar*, edited by Suraj Yengde and Anand Teltumbde, 39–55. Gurgaon: Penguin Random House, 2018.

Hall, Stuart. 'Encoding/Decoding'. In *Culture, Media, Language: Working Papers in Cultural Studies*, edited by Stuart Hall, D. Hobson, A. Lowe, and P. Willis, 4–63. London: Hutchinson, 1980.

Handoo, Jawaharlal. *Folklore: An Introduction*. Mysore: Central Institute of Indian Languages, 1989.

Haraway, Donna J. *Staying with the Trouble: Making Kin in the Chthulucene*. Durham: Duke University Press, 2016.

Harding, Sandra. *Whose Science? Whose Knowledge? Thinking from Women's Lives*. New York: Cornell University Press, 1996.

Hare, Nathan. 'Black Ecology'. *Black Scholar* 1, no. 6 (1970): 2–8.

Harriss-White, Barbara. *India's Market Economy*. Gurgaon: Three Essays, 2005.

Harvey, David. *Justice, Nature and the Geography of Difference*. Cambridge: Blackwell, 1996.

Harvey, Penelope. 'Waste Futures: Infrastructure and Political Experimentation in Southern Peru'. *Ethnos* 82, no. 4 (2017): 672–89.

Hawkins, Gay. *The Ethics of Waste: How We Relate to Rubbish*. Lanham: Rowman and Littlefield, 2006.

Hecht, Gabrielle. 'Interscalar Vehicles for an African Anthropocene: On Waste, Temporality, and Violence'. *Cultural Anthropology* 33, no. 1 (2018): 109–41.

Heidegger, Martin. *Being and Time*. New York: State University of New York Press, 1966.

———. *The Question Concerning Technology and Other Essays*. New York; London: Garland Publishing, 1977.

Hess, Linda. *Bodies of Song: Kabir Oral Traditions and Performative Worlds in North India*. New York: Oxford University Press, 2015.

Heyward, Clare and Dominic Roser, eds. *Climate Justice in a Non-Ideal World*. UK: Oxford University Press, 2016.

Hiltebeitel, Alf. *Rethinking India's Oral and Classical Epics: Draupadi among Rajputs, Muslims, and Dalits*. Chicago: The University of Chicago Press, 1999.

Hird, M. J. 'Knowing Waste: Toward an Inhuman Epistemology'. *Social Epistemology* 26, nos. 3–4 (2012): 453–69.

Hoefe, Rosanne. *Do Leather Workers Matter? Violating Labour Rights and Environmental Norms in India's Leather Production*. Netherlands: Indian Committee of Netherlands, 2017.

Holifield, Ryan, Michael Porter, and Gordon Walker, eds. *Spaces of Environmental Justice*. West Sussex: Wiley-Blackwell, 2010.

Honneth, Axel. *The Struggle for Recognition: The Moral Grammar of Social Conflicts*. Cambridge: MIT Press, 1995.

Horne, Gerald C. *The End of Empires: African Americans and India*. Philadelphia, PA: Temple University Press, 2008.

Hosbey, Justin, Hilda Lloréns, and J. T. Roane. 'Global Black Ecologies'. *Environment and Society: Advances in Research* 13, no. 1 (2022): 1–10.

Ibbetson, Denzil. *Punjab Castes*. Lahore: Superintendent, Government Printing, 1916.

International Dalit Solidarity Network (IDSN). 'A Dalit View on Climate Change'. Accessed on 3 February 2024. https://idsn.org/a-dalit-view-on-climate-change/.

———. 'Upper-caste Farmers Grow Money, We Grow Food'. Accessed on 3 February 2024. https://idsn.org/resources/case-stories/upper-caste-farmers-grow-money-we-grow-food/.

Ikporukpo, Chris O. 'Petroleum, Fiscal Federalism and Environmental Justice in Nigeria'. *Space and Polity* 8, no. 3 (2004): 321–54.

Ilaiah, Kancha. *Post-Hindu India: A Discourse on Dalit-Bahujan, Socio-Spiritual and Scientific Revolution*. Delhi: SAGE, 2009.

Jaaware, Aniket. *Practicing Caste: On Touching and Not Touching*. New York: Fordham University Press, 2019.

Jackson, Iain. 'Politicised Territory: Nek Chand's Rock Garden in Chandigarh'. *GBER* 2, no. 2 (2002): 51–68.

Jadhav, Narendra, ed. *Ambedkar Speaks: 301 Seminal Speeches*. New Delhi: Konark Publications, 2013.

Jairam, Rajani. 'Ecological Concerns in Mahabharata'. *IOSR Journal of Humanities and Social Sciences* 21, no. 5 (May 2016): 63–65.

Jeevan, S. S. *Climate Change Now: The Story of Carbon Colonisation*. New Delhi: Centre for Science and Environment, 2018.

Jha, Mithilesh Kumar. *Language Politics and Public Sphere in North India: Making of the Maithili Movement*. New Delhi: Oxford University Press, 2018.

Jha, V. 'Sustainable Management of Biotic Resources in the Wetlands of North Bihar, India'. In *Aquatic: Conservation, Restoration and Management*, edited by T. V. Ramchandra, N. Ahalya, and C. R. Murthy, 270–77. New Delhi: Capital Publications, 2005.

Jitendra and Rajat Ghai. 'Going Backward'. *Down to Earth*, 31 March 2016. Accessed on 3 February 2024. https://www.downtoearth.org.in/coverage/agriculture/going-backward-53194.

Johnson, Glenn S. 'Environmental Justice: A Brief History and Overview'. In *Environmental Justice in the New Millennium*, edited by Filomina Chioma Steady, 17–46. New York: Palgrave Macmillan, 2009.

Jones, Lindsey. *Overcoming Social Barriers to Adaptation*. London: The Overseas Development Institute, 2010.

Joshi, Chitra. *Lost Worlds: Indian Labour and its Forgotten Histories*. Delhi: Permanent Black, 2003.

Judy, Wajeman. *Feminism Confronts Technology*. London: Polity Press, 1991.

Kalia, Ravi. *Chandigarh: The Making of an Indian City*. Delhi: Oxford University Press, 1987.

Kamath, Anant. '"Untouchable" Cellphones? Old Caste Exclusions and New Digital Divides in Peri-urban Bangalore'. *Critical Asian Studies* 50, no. 3 (2018): 375–94.

Kapoor, S. D. 'B. R. Ambedkar, W. E. B. Du Bois and the Process of Liberation'. *Economic and Political Weekly* 38, nos. 51–52 (2003): 5344–49.

Kapoor, Shivani. 'The Smell of Caste: Leatherwork and Scientific Knowledge in Colonial India'. *South Asia* 44, no. 5 (2021): 983–99.

Kaufman-Osborn, Timothy V. 'Teasing Feminist Sense from Experience'. *Hypatia* 8, no. 2 (1993): 124–44.

Kazi, Rufiat N. and Mangala M. Bote. 'A Cross Sectional Study to Determine the Health Profile of Brick Kiln Workers'. *International Journal of Community Medicine and Public Health* 6, no. 12 (2019): 5135–41.

Kc, Diwas Raja, ed. *Dalit: A Quest for Dignity*. Kathmandu: Nepal Picture Library, 2018.

Kela, Shashank. 'Where the Wild Things Are Not: The Curious Absence of Contemporary Nature Writing in India'. *The Caravan*, 1 April 2018. Accessed on 22 June 2018. http://www.caravanmagazine.in/reviews-essays/absence-contemporary-nature-writing-india.

Khadka, Manohara, Golam Rasul, Lynn Bennett, Shahriar M. Wahid, and Jean-Yves Gerlitz. 'Gender and Social Equity in Climate Change Adaptation in the Koshi Basin: An Analysis for Action'. In *Handbook of Climate Change Adaptation*, edited by Walter Leal Filho, 10–22. Berlin: Springer, 2015.

Khosla, Prabha and Bharati Chaturvedi. 'Mitigation of Greenhouse Gases (CHGs) by Informal Waste Recyclers in Delhi, India'. In *Gender and Climate Change: An Introduction*, edited by Irene Dankelman, 97–99. London: Earthscan, 2010.

Kokkinos, Evgenios and Anastasios I. Zouboulis. 'The Chromium Recovery and Reuse from Tanneries: A Case Study According to the Principles of

Circular Economy'. In *Leather and Footwear Sustainability: Manufacturing, Supply Chain, and Product Level Issues*, edited by Subramanian Senthilkannan Muthu, 123–58. Singapore: Springer, 2020.

Kosambi, D. D. *The Culture and Civilisation of Ancient India in Historical Outline*. Delhi: Vikas Publishing House, 1975.

Kumar, Abhay, Aniruddha Dekha, and Rajat Sinha. *Rural Housing in India: Status and Policy Challenges*. New Delhi: Lokashraya Foundation, 2016.

Kumar, Arun. 'Culture, Development, and Capital of Farce'. In *The Marginalized Self: Tales of Resistance of a Community*, edited by Rahul Ghai, Arvind K. Mishra, and Sanjay Kumar, 29–60. Delhi: Primus Books, 2020.

Kumar, Raj. *Dalit Personal Narratives: Reading Caste, Nation and Identity*. Hyderabad: Orient Blackswan, 2010.

Kumar, Raksha. 'India's Tech Sector Has a Caste Problem'. *Rest of World: Reporting Global Tech Stories*, 19 January 2022. Accessed on 12 September 2022. https://restofworld.org/2022/tech-india-caste-divides/.

Kumar, Sanjay, Arvind Mishra, Badri Narayan, and Rafiul Ahmed. 'Representation, Resistance, and Identity: The Musahars of Middle Gangetic Plain'. In *Interrogating Development: Insights from the Margin*, edited by Frederique Apffel-Marglin, Sanjay Kumar, and Arvind Mishra, 150–71. New Delhi: Oxford University Press, 2010.

Kumari, Sunita. 'Occupational Health of Brick Workers of India'. *International Journal of Health Sciences and Research* 183, no. 8 (September 2018): 183–89.

Kunduri, Eesha. 'Between *Khet* (Field) and Factory, *Gaanv* (Village) and *Sheher* (City): Caste, Gender and the (Re)shaping of Migrant Identities in Urban India'. *Samaj* 19 (2018): 1–19.

Lal, C. K. 'Cultural Flows across a Blurred Boundary'. *South Asian Himal*, 1 February 2002. Accessed on 4 February 2024. https://www.himalmag.com/cultural-flows-across-a-blurred-boundary/.

Latour, Bruno. 'The Anthropocene and the Destruction of the Image of the Globe'. Gifford Lectures on Natural Religion, University of Edinburgh, 18–28 February 2013. Accessed on 4 February 2024. https://www.ed.ac.uk/arts-humanities-soc-sci/news-events/lectures/gifford-lectures/archive/series-2012-2013/bruno-latour.

Leach, Melissa and Robin Mearns, eds. *The Lie of the Land: Challenging Received Wisdom on the African Environment*. Portsmouth: Heinemann, 1996.

Legg, Stephen. *Spaces of Colonialism: Delhi's Urban Governmentalities*. Oxford: Blackwell, 2007.

Limbale, Sharankumar. *The Outcaste: Akkarmashi*. Translated from Marathi by Santosh Bhoomkar. New Delhi: Oxford University Press, 2003.

————. *Towards an Aesthetic of Dalit Literature: History, Controversies and Considerations*. Translated from Marathi by Alok Mukherjee. Hyderabad: Orient Longman, 2004.

Lof, Annette. *More than Meets the Eye? Exploring How Social Constructions Impact Adaptive Capacity to Climate Change*. CTM Stockholm University: Centre for Transdisciplinary Environmental Research, 2006.

Lorea, Carola Erika. *Folklore, Religion and the Songs of a Bengali Madman: A Journey between Performance and the Politics of Cultural Representation*. Leiden: BRILL, 1987.

Lundgren-Kownacki, Karin, Siri M. Kjellberg, Pernille Gooch, Marwa Dabaieh, Latha Anandh, and Vidhya Venugopal. 'Climate Change-Induced Heat Risks for Migrant Populations Working at Brick Kilns in India: A Transdisciplinary Approach'. *International Journal of Biometeorology* 62, no. 10 (2018): 347–58.

MacGregor, Sherilyn. 'Moving beyond Impacts: More Answers to the "Gender and Climate Change" Questions'. In *Understanding Climate Change through Gender Relations*, edited by Susan Buckingham and Virginie Le Masson, 15–30. London: Routledge, 2017.

MacKenzie, Donal and Judy Wajcman. *The Social Shaping of Technology*. London: McGraw Hill Education, 1999.

Mahalingam, Ramaswami, Srinath Jagannathan, and Patturaja Selvaraj. 'Decaticization, Dignity, and "Dirty Work" at the Intersections of Caste, Memory, and Disaster'. *Business Ethics Quarterly* 29, no. 2 (April 2019): 213–39.

Mallabarman, Adwaita. *A River Called Titash*. Translated from Bengali by Kalpana Bardhan. Berkeley: University of California Press, 1993.

Mangala, Pradnya. 'Climate Justice in India: A Critical Overview'. *Round Table India*, 6 October 2019. Accessed on 4 February 2024. https://roundtableindia. co.in/index.php?option=com_content&view=article&id=9734:climate-justice-in-india-a-critical-overview&catid=119:feature&Itemid=132.

'Maun', Prafulla Kumar Singh and Ashwini Kumar Alok, eds. *Dina–Bhadri: Musaharon ki Samagra Sanskriti*. New Delhi: Samyak Prakashan, 2015.

Martin, Adrian, 'A River Called Titas: River of No Return'. *The Criterion Collection*, 12 December 2013. Accessed on 4 February 2024. https://www.criterion.com/current/posts/2990-a-river-called-titas-river-of-no-return.

Martinez-Alier, Joan, Leah Temper, Daniela Del Bene, and Arnim Scheidel. 'Is There a Global Environmental Justice Movement?' *Journal of Peasant Studies* 43, no. 3 (April 2016): 731–55.

Marx, Karl. *Capital*. Vol. 1. Moscow: Progress Publishers, 1975.

————. *The Economic and Philosophical Manuscripts of 1844*. Amherst: Prometheus, 1988 (1927).

Marx, Karl and Frederick Engels. *Selected Works.* Vol. 1. Moscow: Progress Publishers, 1978.

Massey, James. *Dalit Theology: History, Context, Text and Whole Salvation.* New Delhi: Manohar Publications, 2014.

Mayberry, Maralee, Banu Subramaniam, and Lisa H. Weasel, eds. *Feminist Science Studies: A New Generation.* London; New York: Routledge, 2001.

Mayer, Sylvia, ed. *Restoring the Connection to the Natural World: Essays on the African American Environmental Imagination.* New Brunswick: Transaction Publishers, 2003.

Mbembe, Achille. *Critique of Black Reason.* Translated by Laurent Dubois. Durham, NC: Duke University Press, 2017.

McCright, Aaron M. and Riley E. Dunlap. 'Cool Dudes: The Denial of Climate Change among Conservative White Males in the United States'. *Global Environmental Change* 21, no. 4 (2011): 1163–72.

McDonald, David A. 'Environmental Racism and Neoliberal Disorder in South Africa'. In *The Quest for Environmental Justice: Human Rights and the Politics of Pollution*, edited by Robert D. Bullard, 255–78. San Francisco: Sierra Club Books, 2005.

Mehta, V. P. and Nek Chand Saini. *Rock Garden: A Vision of Creativity.* Chandigarh: Arun Publishing, 2010.

Menon, Sunil. 'The Global Dalit, The Indian Black: Cornel West in Conversation with Suraj Yengde'. *Outlook India*, 3 October 2020. Accessed 12 October 2022. https://www.outlookindia.com/magazine/story/india-news-independence-day-special-whos-indias-george-floyd-heres-exposing-a-racist-india/303552.

Michel, David and Amit Pandya, eds. *Indian Climate Policy: Choices and Challenges.* Washington, DC: The Henry L. Stimson Center, 2009.

Mirza, Shireen. 'Figure of the Halalkhore: Caste and Stigmatised Labour in Colonial Bombay'. *Economic and Political Weekly* 53, no. 31 (4 August 2018): 79–85.

Mitter, Swasti and Sheila Rowbotham. *Women Encounter Technology: Changing Patterns of Employment in the Third World.* London; New York: Routledge, 1995.

Modi, Narendra. 'PM's Statement at the United Nations Summit for the Adoption of Post-2015 Development Agenda'. *Narendra Modi*, 25 September 2015. Accessed 3 February 2024. https://www.narendramodi.in/text-of-pm-s-statement-at-the-united-nations-summit-for-the-adoption-of-post-2015-development-agenda-332923.

Moizels, John. 'Nek Chand: Creator of a Magical World'. In *Vernacular Visionaries, International Outsider Art*, edited by A. Carlano, 66–77. New Haven: Yale University Press, 2003.

————. *Raw Creation*. New York: Phaidon Press, 1996.

Moon, Vasant. *Growing Up Untouchable in India: A Dalit Autobiography*. Translated from Marathi by Gail Omvedt. Maryland: Rowman & Littlefield Publishers, 2001.

Moore, Donald S., Jake Kosek, and Anand Pandian, eds. *Race, Nature, and the Politics of Difference*. Durham: Duke University Press, 2003.

Moosvi, Shireen. 'Environmental Concerns in Mughal Era'. *Journal of History and Social Sciences* 1, no. 1 (July–December 2010): 1–4.

Morrison, Kathleen D., 'Provincializing the Anthropocene'. *Seminar*. Accessed 4 February 2024. https://www.india-seminar.com/2015/673/673_kathleen_morrison.htm.

Moss, Jeremy, ed. *Climate Change and Social Justice*. Victoria: Melbourne University Press, 2009.

Mosse, David. 'The Modernity of Caste and the Market Economy'. *Modern Asian Studies* 54, no. 4 (2020): 1225–71.

————. *The Rule of Water: Statecraft, Ecology, and Collective Action in South India*. Delhi: Oxford University Press, 2003.

Mukherjee, Rudrangshu. *Spectre of Violence: The 1857 Kanpur Massacres*. Delhi: Viking, 1998.

Mukherjee, Upamanyu Pablo. *Post-colonial Environments: Nature, Culture and the Contemporary Indian Novel in English*. New York: Palgrave, 2010.

Mukul. *Steel Bartan Udhog: Maut Se Jooghte Mazdoor*. Delhi: Delhi General Mazdoor Front, 1989.

Mukul. 'Steel Workers of Delhi: At the Mercy of the Owner'. *Forum Gazette* (1989): 8–9.

Muthu, Subramanian Senthilkannan, ed. *Leather and Footwear Sustainability: Manufacturing, Supply Chain, and Product Level Issues*. Singapore: Springer, 2020.

Myrdal, Gunnar. *An Inquiry into the Poverty of Nations*. New York: Twentieth Century Fund, 1968.

Nagraj, D. R. *The Flaming Feet and Other Essays: The Dalit Movement in India*. Ranikhet: Permanent Black, 2010.

Nair, L. R. *Why Chandigarh?* Simla: Punjab Government, 1950.

Namishray, Mohan Dass. *Caste and Race: A Comparative Study of B. R. Ambedkar and Martin Luther King*. Jaipur: Rawat Publications, 2003.

Narain, Sunita, Prodipto Ghosh, N. C. Saxena, Jyoti Parikh, and Preeti Soni. *Climate Change: Perspectives from India*. Delhi: UNDP India, 2009.

Narayan, Badri. 'Myth, Culture, and Democracy'. In *The Marginalized Self: Tales of Resistance of a Community*, edited by Rahul Ghai, Arvind K. Mishra and Sanjay Kumar, 61–81. Delhi: Primus Books, 2020.

————. *The Making of the Dalit Public in North India, Uttar Pradesh, 1950–Present*. New Delhi: Oxford University Press, 2011.

National Dalit Watch of National Campaign on Dalit Human Rights and Society for Promotion of Wastelands Development. *Impact of Climate Change on Life and Livelihood of Dalits: An Exploratory Study from Disaster Risk Reduction Lens*. Delhi: National Dalit Watch of National Campaign on Dalit Human Rights and Society for Promotion of Wastelands Development, 2013.

National Dalit Watch of National Campaign on Dalit Human Rights. *Addressing Caste Discrimination in Humanitarian Response*. New Delhi: NCDHR, 2011.

National Population and Housing Census 2011. Kathmandu: Central Bureau of Statistics, Government of Nepal, 2012.

'National Seminar on Dalit Studies and Higher Education: Exploring Content Material for a New Discipline'. Deshkal Society, Delhi, 28 February 2004.

Natrajan, Balmurli and Paul Greenough, eds. *Against Stigma: Studies in Caste, Race and Justice since Durban*. Hyderabad, India: Orient Blackswan, 2009.

Nepal Academy. *Dina–Bhadri: Sachitra Sankalan*. Kathmandu, 2014.

Nielsen, Kenneth Bo, Siddharth Sareen, and Patrik Oskarsson. 'The Politics of Caste in India's New Land Wars'. *Journal of Contemporary Asia* 50, no. 5 (2020): 684–95.

Ninan, K. N. 'Climate Change and Rural Poverty Levels in India'. *Economic and Political Weekly* 54, no. 2 (January 2019): 36–43.

Nishime, Leilani and Kim D. Hester Williams, eds. *Racial Ecologies*. Seattle: University of Washington Press, 2018.

Noor, Poppy. 'Detroit Families Still Without Clean Water Despite Shutoffs Being Lifted'. *The Guardian*, 20 May 2020. Accessed on 3 February 2024. https://bit.ly/34GJPQI.

O'Hanlon, Rosalind. *Caste, Conflict, and Ideology: Mahatma Jotirao Phule and Low-caste Protest in Nineteenth-Century Western India*. Ranikhet: Permanent Black, 2002.

Omvedt, Gail. 'Why Dalits Dislike Environmentalists'. *The Hindu*, 25 June 1997.

Onta, Nisha and Bernadette P. Resurreccion. 'The Role of Gender and Caste in Climate Adaptation Strategies in Nepal: Emerging Change and Persistent Inequalities in the Far-Western Region'. *Mountain Research and Development* 31, no. 4 (2011): 351–56.

Ory, Ferene Gyula, ed. *Strategies and Methods to Promote Occupational Health: Industrial Counselling in Tanneries in India*. Les Pailles, Mauritius: Preci-ex Limited, 1997.

Ory, F. G., F. U. Rahman, V. Katagade, A. Shukla, and A. Burdorf. 'Respiratory Disorders, Skin Complaints and Low-Back Problems among Tannery

Workers in Kanpur, India'. In *Strategies and Methods to Promote Occupational Health in Low-Income Countries: Industrial Counselling in Tanneries in India*, edited by Ferene Gyula Ory, 740–46. Les Pailles, Mauritius: Preci-ex Limited, 1997.

Ottinger, Gwen and Benjamin R. Cohen, eds. *Technoscience and Environmental Justice*. Cambridge, Massachusetts: MIT Press, 2011.

Paik, Shailaja. 'Building Bridges: Articulating Dalit and African American Women's Solidarity'. *WSQ: Women's Studies Quarterly* 42, nos. 3–4 (2014): 74–96.

Pal, Sumedha. 'Overlooked Correlation between Climate Change and Social Exclusion'. *NewsClick*, 11 July 2019. Accessed 3 February 2024. https://www.newsclick.in/overlooked-correlation-climate-change-social-exclusion.

Pandey, Gyanendra. *A History of Prejudice: Race, Caste, and Difference in India and the United States*. Cambridge: Cambridge University Press, 2013.

———. 'Politics of Difference: Reflections on Dalit and African American Struggles'. *Economic and Political Weekly* 45, no. 19 (2010): 62–69.

Pandey, Priyanka and Sandeep Pandey. 'Survey at an IIT Campus Shows How Caste Affects Students' Perceptions'. *Economic and Political Weekly* 53, no. 9 (March 2018): 1–8.

Pandian, Anand. *Crooked Stalks: Cultivating Virtue in South India*. Durham: Duke University Press, 2009.

Pandian, M. S. S. 'Writing Ordinary Lives'. *Economic and Political Weekly* 43, no. 38 (20 September 2008): 35–40.

Parikh, Sunita. *The Politics of Preference: Democratic Institutions and Affirmative Action in the United States and India*. Ann Arbor, MI: University of Michigan Press, 1997.

Pattanaik, D. R., M. Mohapatra, A. K. Srivastava, and Arun Kumar. "Heat Wave over India during Summer 2015: An Assessment of Real Time Extended Range Forecast'. *Meterology and Atmospheric Physics* 129 (August 2017): 375–93.

Pawar, Urmila. *The Weave of My Life: A Dalit Woman's Memoirs*. Translated from Marathi by Maya Pandit. New York: Columbia University Press, 2009.

Peiry, Lucienne and Philippe Lespinasse. *Nek Chand's Outsider Art: The Rock Garden of Chandigarh*. Paris: Flammarion, 2015.

Pfaffenberger, Bryan. 'Social Anthropology of Technology'. *Annual Review of Anthropology* 21, no. 1 (1992): 491–516.

Pokhrel, Bhabani. 'Strained Identity: Cultural and Religious Rituals of a Musahar Community'. *Social Inquiry* 2, no. 1 (2020): 128–50.

Practical Action. *Technology Action: A Call to Action*. Warwickshire, UK: Practical Action, 2016.

Prakash, Aseem. *Dalit Capital: State, Markets and Civil Society in Urban India*. New Delhi: Routledge, 2015.

Prakash, Brahma. *Cultural Labour: Conceptualizing the 'Folk Performance' in India*. New Delhi: Oxford University Press, 2019.

Prakash, Pranav. 'The Hidden Casteism of Climate Change Reporting in India'. *The Quint*, 27 October 2016. Accessed on 3 February 2024. https://www.thequint.com/news/environment/the-hidden-casteism-in-climate-change-media-reporting-in-india-dalit-agriculture.

Prakash, Vikramaditya. *Chandigarh's Le Corbusier: The Struggle for Modernity in Postcolonial India*. Ahmedabad: Mapin Publishing, 2002.

Prashad, Vijay. 'Afro-Dalits of the Earth, Unite!'. *African Studies Review* 43, no. 1 (April 2000): 189–201.

Prayer, Mario. 'Freedom in the River: Bengali *Bhadralok* Consciousness in Manik Bandopadhyay's *Padmanadir Majhi*'. In *The Human Person and Nature in Classical and Modern India*, edited by Fabrizio Serra, 165–80. Rome: Sapienza, 2015.

Raheja, Gloria Goodwin and Ann Grodzins Gold. *Listen to the Heron's Words*. Berkeley: University of California Press, 1994.

Rahman, Abdur, ed. *Science and Technology in Indian Culture: A Historical Perspective*. Delhi: NISTDS, 1984.

Rajan, S. Ravi. 'Environmental Justice in India'. *Environmental Justice* 7, no. 5 (2014): 117–21.

Rajashekar, V. T. *Dalit: The Black Untouchables of India*. Atlanta, GA: Clarity Press, 1995.

Rajivlochan, Meeta, Kavita Sharma, and Chitleen K. Sethi. *Chandigarh Lifescape: Brief Social History of a Planned City*. Chandigarh: Chandigarh Administration, 1999.

Rakesh, Ram Dayal. *Folk Tales from Mithila*. New Delhi: Nirala Publications, 1996.

Ram, Mahendra Narayan. *Panchalok Devta*. Patna: Bihar Hindi Granth Academy, 2003.

Ram, Mahendra Narayan and Phulo Paswan, eds. *Dina–Bhadri Lokgatha*. Delhi: Sahitya Academy, 2012.

Rammohan, K. T. *Tales of Rice: Kuttanad, Southwest India*. Thiruvananthapuram: Centre for Development Studies, 2006.

Ranco, Darren and Dean Suagee. 'Tribal Sovereignty and the Problem of Difference in Environmental Regulation: Observations on "Measured Separatism" in Indian Country'. *Antipode* 39, no. 4 (2007): 691–707.

Ranganathan, Malini. 'Caste, and the Making of Environmental Unfreedoms in Urban India'. *Ethnic and Racial Studies* 45, no. 2 (2022): 257–77.

Rangarajan, Swarnalatha. *Ecocriticism: Big Ideas and Practical Strategies*. Hyderabad: Orient Blackswan, 2018.

Rao, Anupama. *The Caste Question: Dalits and the Politics of Modern India*. Ranikhet: Permanent Black, 2009.

Rao, Velcheru Narayana. *Text and Tradition in South India*. Albany: SUNY Press, 2016.

Rashtriya Garima Abhiyan (National Dignity Campaign). 'Women Liberation March'. In Hindi, 2013. Accessed on 3 February 2024. http://idsn.org/wpcontent/uploads/user_folder/pdf/New_files/Key_Issues/Manual_scavenging/Maila_Mukti_Yatra_2012-13_-_Note.pdf.

Rath, Sharadini. 'A Journey through Madhubani'. 2015. Accessed on 13 March 2022. https://www.phalanx.in/pdf/dini.pdf.

Ravikumar. *Venomous Touch: Notes on Caste, Culture and Politics*. Kolkata: Samya, 2017.

Rawat, Ramnarayan S. *Reconsidering Untouchability: Chamars and Dalit History in North India*. Ranikhet: Permanent Black, 2012.

Rege, Sharmila. 'Conceptualising Popular Culture: "Lavani" and "Powada" in Maharashtra'. *Economic and Political Weekly* 37, no. 11 (2002): 1038–47.

———. *Writing Caste/Writing Gender: Narrating Dalit Women's Testimonies*. New Delhi: Zubaan, 2006.

Reno, Joshua. 'Waste and Waste Management'. *Annual Review of Anthropology* 44, no. 1 (October 2015): 557–72.

Roane, J. T. 'Plotting the Black Commons'. *Souls* 20, no. 3 (2018): 239–66.

Robb, Peter, ed. *The Concept of Race in South Asia: Understanding and Perspectives*. Delhi: Oxford University Press, 1995.

Roberts, J. Timmons and Nikki Demetria Thanos. *Trouble in Paradise: Globalization and Environmental Crisis in Latin America*. New York: Routledge, 2003.

Rodrigues, Valerian, ed. *The Essential Writings of B. R. Ambedkar*. Delhi: Oxford University Press, 2002.

Roghair, Gene H. *The Epic of Palnadu. A Study and Translation of Palnti Virula Katha, a Telugu Oral Tradition from Andhra Pradesh, India*. Oxford: Clarendon Press, 1982.

Roy, Indrajit. 'Emancipation as Social Equality: Subaltern Politics in Contemporary India'. *Focaal* 76 (December 2016): 15–30.

———. 'Utopia in Crisis? Subaltern Imaginations in Contemporary Bihar'. *Journal of Contemporary Asia* 45, no. 4 (2015): 640–59.

Roy, Kartik C. and Cal Clark, eds. *Technological Change and Rural Development in Poor Countries: Neglected Issues*. Calcutta: Oxford University Press, 1994.

Ruffin, Kimberly N. *Black on Earth: African American Ecoliterary Traditions*. Athens: The University of Georgia Press, 2010.

SAARC Energy Centre. *Evaluating Energy Conservation Potential of Brick Production in India*. Islamabad: SAARC Energy Centre, 2013.

Saikia, Arupjyoti. *The Unquiet River: A Biography of the Brahmaputra*. Delhi: Oxford University Press, 2019.

Saini, Anuj. 'My Father's Kingdom of Gods and Goddesses'. In *Chandigarh: An Anthology*, edited by A. Uberoi, 42–50. Chennai: Creative Workshop, 2021.

Sandler, Ronald and Phaedra C. Pezzullo, eds. *Environmental Justice and Environmentalism: The Social Justice Challenge to the Environmental Movement*. Cambridge, MA: MIT Press, 2007.

Sanitation Workers Movement. 'Bhim March Daily Reports'. In Hindi, 2016. Accessed on 3 February 2024. https://www.safaikarmachariandolan.org/resources.

Sarin, Madhu. *Urban Planning in the Third World: The Chandigarh Experience*. New York: Routledge, 2019.

Sarukkai, Sundar. 'Experience and Theory: From Habermas to Gopal Guru'. In *The Cracked Mirror: An Indian Debate on Experience and Theory*, edited by Gopal Guru and Sundar Sarukkai, 29–45. New Delhi: Oxford University Press, 2012.

Satiroglu, Irge and Narae Choi. *Development-Induced Displacement and Resettlement: New Perspectives on Persisting Problems*. London: Routledge, 2017.

Satyanarayana, K. and Susie Tharu, eds. *No Alphabet in Sight: New Dalit Writing from South India, Dossier 1: Tamil and Malayalam*. New Delhi: Penguin Books, 2011.

————, eds. and intro. *Steel Nibs Are Sprouting: New Dalit Writing from South India, Dossier II: Kannada and Telugu*. Noida: Harper Collins, 2013.

Saxena, K. B. *Prevention of Atrocities against Scheduled Castes: Policy and Performance*. Delhi: National Human Rights Commission, 2004.

Schlosberg, David. *Defining Environmental Justice: Theories, Movements, and Nature*. Oxford: Oxford University Press, 2007.

————. 'The Justice of Environmental Justice: Reconciling Equity, Recognition, and Participation in a Political Movement'. In *Moral and Political Reasoning in Environmental Practice*, edited by Andrew Light and Avner De-Shalit, 77–106. Cambridge, MA: MIT Press, 2003.

Schumacher, E. F. *Good Work*. London: Abacus, 1973.

————. *Small Is Beautiful: A Study of Economics as If People Mattered*. London: Blond and Briggs, 1973.

Seabrook, Jeremy. *Class, Caste and Hierarchies*. Jaipur: Rawat Publications, 2005.

Shah, Esha. *Social Designs: Tank Irrigation Technology and Agrarian Transformation in Karnataka, South India*. Delhi: Orient Longman, 2003.

————. 'Telling Otherwise: A Historical Anthropology of Tank Irrigation Technology in South India'. *Technology and Culture* 49, no. 3 (July 2008): 652–74.

Shakti. *A Roadmap for Cleaner Brick Production in India*. Delhi: Shakti Sustainable Energy Foundation, 2012.

Shankar, S. and Charu Gupta. *Caste and Life Narratives*. Delhi: Primus Books, 2019.

Sharan, Awadhendra. *In the City out of Place: Nuisance, Pollution and Dwelling in Delhi*. New Delhi: Oxford University Press, 2014.

Sharma, Mukul. *Caste and Nature: Dalits and Indian Environmental Politics*. New Delhi: Oxford University Press, 2017.

————. 'Dalits and Indian Environmental Politics'. *Economic and Political Weekly* 47, no. 23 (2012): 46–52.

————. *Dalit aur Prakriti: Jati aur Paryavaran Aandolan*. New Delhi: Vani Prakashan, 2020.

————. 'Deena–Bhadri's Sacrifice Is Ours: Everyday Life of Musahars in North Bihar'. *Labour File* 4, nos. 5–6 (May–June 1998): 3–25.

————. 'The Untouchable Present: Everyday Life of Musahars in North Bihar'. In *Village Society*, edited by Surinder S. Jodhka, 93–102. New Delhi: Orient BlackSwan, 2012.

Sharma, Mukul and Ashok Bharti. 'In Lieu of an Introduction'. In *Defining Dignity: An Anthology of Dreams, Hopes and Struggles*, edited by Mukul Sharma and Sana Das, 2–12. Delhi: World Dignity Forum, 2005.

Sharma, Pradeep K. *Dalit Politics and Literature*. Delhi: Shipra, 2006.

Sharma, Sangeet. *The Corb's Capital: Journey through Chandigarh Architecture*. Chandigarh: A3 Foundation, 2014.

Sheller, Mimi. *Mobility Justice: The Politics of Movement in an Age of Extremes*. London: Verso, 2018.

Shohat, Ella. 'Notes on the Post-Colonial'. *Social Text* 31/32 (1992).

Shrestha, Nanda R. and Dennis Conway. 'Ecopolitical Battles at the Terai Frontier of Nepal: An Emerging Human and Environmental Crisis'. *International Journal of Population Geography* 2, no. 4 (December 1996): 313–31.

Shukla, Abhay, Satish Kumar, and F. G. Ory. 'Occupational Health and the Environment in an Urban Slum in India'. *Social Science Medicine* 33, no. 5 (1991): 597–603.

Siddalingaiah. *Ooru Keri*. Translated from Kannada by S. R. Ramakrishna. New Delhi: Sahitya Akademi, 2003.

Sinclair, Bruce. *Technology and the African-American Experience: Needs and Opportunities for Study*. Cambridge, Massachusetts: The MIT Press, 2004.

Singh, D. P. 'Women Workers in the Brick Kiln Industry in Haryana, India'. *Indian Journal of Gender Studies* 12, no. 1 (2005): 83–96.

Singh, K. S. 'Musahar: Community, Context and Equality'. In *Asserting Voices: Changing Culture, Identity and Livelihood of the Musahars in the Gangetic Plains*, edited by Hemant Joshi and Sanjay Kumar, 134–42. Delhi: Deshkal Publication, 2002.

Singh, Virendra Kumar. *Hamare Lok Devi-Devta*. Delhi: Samiksha, 1999.

Sinha, Frances, K. A. Srinivasan, Rajiv Kumar Singh, and Viji Srinivasan. *The Blue Revolution: Case-Study of Women in the Inland Fisheries Sector*. Delhi: Har-Anand Publications, 1994.

Sircar, Srilata. 'Reimagining Climate Justice as Caste Justice'. In *Climate Justice in India*, edited by Prakash Kashwan), 162–81. UK: Cambridge University Press, 2022.

Skillington, Tracey. *Climate Justice and Human Rights*. New York: Palgrave Macmillan, 2017.

Slate, Nico, ed. *Black Power Beyond Borders: The Global Dimensions of the Black Power Movement*. New York: Palgrave Macmillan, 2012.

Slovic, Scott. 'Seasick among the Waves of Ecocriticism: An Inquiry into Alternative Historiographic Metaphors'. In *Environmental Humanities: Voices from the Anthropocene*, edited by Serpil Oppermann and Serenella Iovino, 99–111. London: Rowman and Littlefield International, 2016.

Slovic, Scott, Swarnalatha Rangarajan, and Vidya Sarveswaran, eds. *Ecocriticism of the Global South*. Lanham: Lexington Books, 2015.

Smith, A. 'The Alternative Technology Movement: An Analysis of its Framing and Negotiation of Technology Development'. *Human Ecology Review* 12, no. 2 (2005): 106–19.

Smith, John D. *The Epic of Pabuji: A Study, Transcription and Translation*. Cambridge: Cambridge University Press, 1991.

Smith, Kimberly K. *African American Environmental Thought: Foundations*. Kansas: University Press of Kansas, 2007.

Somanathan, E. and Rohini Somanathan. 'Climate Change: Challenges Facing India's Poor'. *Economic and Political Weekly* 44, no. 31 (August 2009): 51–58.

Spivak, Gayatri Chakravorty. 'City, Country, Agency'. *Future Anterior* 16, no. 2 (2019): 59–85.

Spronk, Susan. 'Covid-19 and Structural Inequalities: Class, Gender, Race and Water Justice'. In *Public Water and Covid-19: Dark Clouds and Silver Linings*, edited by David A. McDonald, Susan J. Spronk, and Daniel Chavez, 25–48. Kingston: Municipal Services Project, 2020.

Stam, Robert. 'Beyond Third Cinema: The Aesthetics of Hybridity'. In *Rethinking Third Cinema*, edited by A. R. Guneratne and W. Dissanayake, 31–48. New York: Routledge, 2003.

Steady, Filomina Chioma, ed. *Environmental Justice in the New Millennium*. New York: Palgrave Macmillan, 2009.

Stevens, Lara, Peta Tait, and Denise Varney, eds. *Feminist Ecologies: Changing Environments in the Anthropocene*. Switzerland: Palgrave Macmillan, 2018.

Stewart, Mart A. 'Slavery and the Origins of African American Environmentalism'. In *'To Love the Wind and the Rain': African Americans and Environmental History*, edited by Dianne D. Glave and Mark Stoll, 9–20. Pittsburgh: University of Pittsburgh Press, 2006.

Subha, Bhim. *Himalayan Waters: Promise and Potential, Problems and Politics*. Kathmandu: PANOS South Asia, 2001.

Sukumar, Arun Mohan. *Midnight's Machines: A Political History of Technology in India*. Haryana: Penguin, 2019.

Sundberg, Juanita. 'Tracing Race: Mapping Environmental Formations in Environmental Justice Research in Latin America'. In *Environmental Justice in Latin America*, edited by David V. Carruthers, 25–-48. Cambridge: MIT Press, 2008.

Sur, Abha. *Dispersed Radiance: Caste, Gender, and Modern Science in India*. Delhi: Navayana, 2011.

Swaroop, Kanthi. 'Manual Scavenging and Technology: Notes on Indian Urbanism'. Seminar Series, Indian Institute of Technology, Gandhinagar, 25 April 2022.

Swati Rajput, Kavita Arora, and Rachna Mathur. *Urban Green Space, Health Economics and Air Pollution in Delhi*. London: Routledge, 2021.

Sze, Julie. 'From Environmental Justice Literature to the Literature of Environmental Justice'. In *The Environmental Justice Reader: Politics, Poetics and Pedagogy*, edited by Joni Adamson, Mei Mei Evans, and Rachel Stein, 163–80. Arizona: University of Arizona Press, 2006.

Talib, Mohammad. *Writing Labour: Stone Quarry Workers in Delhi*. New Delhi: Oxford University Press, 2010.

Tandale, Dadasaheb. 'Caste, Economic Inequality and Climate Justice in India'. In *Human Rights and Economic Inequalities*, edited by Gillian Macnaughton, Diane F. Frey, and Catherine F. Frey, 217–44. Cambridge: Cambridge University Press, 2021.

Taylor, Dorceta E. *Race, Class, Gender, and American Environmentalism*. Portland: US Department of Agriculture, 2002.

———. *Toxic Communities: Environmental Racism, Industrial Pollution, and Residential Mobility*. New York: New York University Press, 2014.

Thorat, Sukhadeo, Paul Attewell, and Firdaus Fatima Rizvi. *Urban Labour Market Discrimination*. Vol. 3, no. 1. Delhi: Indian Institute of Dalit Studies, 2009.

Todd, Zoe. 'Indigenizing the Anthropocene'. In *Art in the Anthropocene: Encounters among Aesthetics, Politics, Environments and Epistemologies*, edited by Heather Davis and Etienne Turpin, 241–54. London: Open Humanities Press, 2015.

Turner, Victor. *The Anthropology of Performance*. New York: PAJ Publications, 1988.

Twain, Mark. *Following the Equator and Anti-imperialist Essays*. New York: Oxford University Press, 1996.

Umberger, Leslie. *Nek Chand: Healing Properties*. Wisconsin: John Michael Kohler Arts Center, 2000.

Upadhyay, Nilay. *Pahar*. New Delhi: Radhakrishan Prakashan, 2015.

Vaidya, Anand. 'This Forests Belong to Us'. In Hindi, *Seminar*, 2017. Accessed on 3 February 2024. https://www.india-seminar.com/2017/690/690_anand_vaidya.htm.

Valmiki, Om Prakash. *Joothan: A Dalit's Life*. Translated from Hindi by Arun Prabha Mukherjee. Kolkata: Samya, 2007.

Vani, M. S., Rohit Asthana, Narayan Belbase, L. B. Thapa, Patricia Moore, and Firuza Pastakia. *Environmental Justice and Rural Communities: Studies from India and Nepal*. Bangkok: IUCN, 2007.

Visweswaran, Kamala. *Un/common Cultures: Racism and the Rearticulation of Cultural Difference*. Durham, NC: Duke University Press, 2010.

Vital, Anthony. 'Toward an African Ecocriticism: Postcolonialism, Ecology and Life and Times of Michael K'. *Research in African Literatures* 39 (Spring 2008): 87–121.

Wadley, Susan Snow. *Raja Nal and the Goddess: The North Indian Epic Dhola in Performance*. Bloomington: Indiana University Press, 2004.

Wainwright, Joel. 'The Geographies of Political Ecology: After Edward Said'. *Environment and Planning* 37, no. 6 (June 2005).

Wakankar, Milind. *Subalternity and Religion: The Prehistory of Dalit Empowerment in South Asia*. New York: Routledge, 2010.

Wells, Christopher W., ed. *Environmental Justice in Postwar America: A Documentary Reader*. Seattle: University of Washington Press, 2018.

White, Rob. *Environmental Harm: An Eco-justice Perspective*. Bristol: Polity Press, 2013.

Wilkerson, Isabel. *Caste: The Origins of Our Discontents*. New York: Random House, 2020.

Williams, Raymond. *The Sociology of Culture*. Chicago: University of Chicago Press, 1981.

Wilson, Bezwada and Bhasha Singh. *The Long March to Eliminate Manual Scavenging: India Exclusion Report 2016*. Delhi: Yoda Press, 2016, 298–319.

Yadav, Shiv Prasad, ed. *Maithili Dalit Lokgatha ao Sanskriti*. Delhi: Sahitya Academy, 2015.

Yengde, Suraj and Anand Teltumbde, eds. *The Radical in Ambedkar: Critical Reflections*. Gurgaon: Penguin Random House, 2018.

Yusoff, Kathryn. *A Billion Black Anthropocenes or None*. Minneapolis: University of Minnesota Press, 2018.

Zalasiewicz, Jan A. *The Earth After Us*. Oxford: Oxford University Press, 2008.

Zelliot, Eleanor. *From Untouchable to Dalit: Essays on the Ambedkar Movement*. New Delhi: Manohar, 1992.

Index

carbon footprints, 187
carbon monoxide (CO), 198
caste
  consciousness, 86, 110
  discrimination, 122, 203
  hierarchy, 22, 151, 164–65, 181, 191, 225
  identity, 141, 162, 169, 178, 190
  labour, 12, 133, 142, 171
  stereotypes, 22, 190
  supremacy, 174
  worker, 166, 170
caste-based
  atrocities, 24
  discrimination, 104, 133, 139, 165
    Indian experiences in disaster response and DRR, 205
  local economy, 200
  occupational hierarchy, 12
  rituals and symbols, 59
  subjugation, 13
  technology, 13
caste–capitalist economy, 14
caste–colonial–capitalist systems, 107
caste–corporatist capitalism, 201
caste-determined economic organization, 156
caste divisions, in labour, 157, 166, 188
caste economy, 21, 187
  and climate justice, 200–02
caste–industry–technology nexus, 13
caste justice, discourse on, 30
caste–occupation nexus, 201
caste society, characteristic of, 141
caste–technology relations, 13
casual labourers, 189
Census of India (2011), 189
Central Pollution Control Board, 211n42

Centre of Indian Trade Unions, 155, 156
Chakrabarty, Dipesh, 110, 131, 175
Chamars, 162, 224
  'Chamarisation' of leatherwork, 171
  colonial association of, 170
  issues of rights and dignity, 224
  leatherworking schools, 170
  Nara-Maveshi movement of the 1950s–1960s, 224
  'natural' association of, 170
  occupational image of, 169
  tanneries and working conditions of, 170
Chandigarh city
  accelerating India into the Modern Age, 112–18
  building project, 121
  as capital of Punjab and Haryana, 112
  Capitol Complex, 115, 121
  Capitol Project, 109–18
  Rock Garden, 9–10, 21, 107, 109
Chand, Nek, 9–10, 16–17, 21, 104, 107–08, 118, 121, 132
  archives of Anthropocene, 124
  birth of, 107
  creation of the Rock Garden in Chandigarh city, 9–10, 21, 107, 109
  as government employee in PWD, 109
  life narratives, 118–24
  Mali caste, 108, 122
  migration to India, 107
  rock collection, 123
  Rock Garden 'Allegories of the Anthropocene', 126
chaturvarnya system, 229